T0211496

Lecture Notes in Computer Science   10268

Commenced Publication in 1973
Founding and Former Series Editors:
Gerhard Goos, Juris Hartmanis, and Jan van Leeuwen

More information about this series at http://www.springer.com/series/7409

Enrique Alba · Francisco Chicano
Gabriel Luque (Eds.)

Smart Cities

Second International Conference, Smart-CT 2017
Málaga, Spain, June 14–16, 2017
Proceedings

Editors
Enrique Alba
University of Malaga
Malaga
Spain

Francisco Chicano
University of Malaga
Malaga
Spain

Gabriel Luque
University of Malaga
Malaga
Spain

ISSN 0302-9743   ISSN 1611-3349   (electronic)
Lecture Notes in Computer Science
ISBN 978-3-319-59512-2   ISBN 978-3-319-59513-9   (eBook)
DOI 10.1007/978-3-319-59513-9

Library of Congress Control Number: 2017942983

LNCS Sublibrary: SL3 – Information Systems and Applications, incl. Internet/Web, and HCI

Printed on acid-free paper

This Springer imprint is published by Springer Nature
The registered company is Springer International Publishing AG
The registered company address is: Gewerbestrasse 11, 6330 Cham, Switzerland

Preface

We are happy to present the proceedings of the Second International Conference on Smart Cities held in Málaga, Spain, during June 14–16, 2017. The subject of smart cities is a growing field that has still not reached the whole realm of science, nor many other domains like municipal management, politics, and economy. Research in smart cities is undoubtedly a very important topic, but a young one from a scientific point of view. It is true that many articles are starting to appear in a few other similar conferences and journals. However, the field still needs some time to become established and recognized as a separate domain of study.

Indeed, some of the open issues are very basic: What is the meaning of *smart* in smart city? Is any single paper on, e.g., routes between two points a smart city paper? The number of open questions is large. However, a few things are clear: There are many domains in the city (mobility, energy, construction, citizens, social implications, economy, technology, tourism) and the studies on smart cities should have a corresponding holistic vision in their contents. Idle studies, typical in research to date, need to grow to consider the city and several of its aspects simultaneously. As to the *smart* part, well, who knows what is smart at all? Currently, there is only a fuzzy definition, and again we need to go beyond the mere use of a sensor or a smartphone to make applications *intelligent*.

With this conference, we hope to advance in these and other questions. Our expectations are great in the sense that, at the very least, we are trying to pose the correct questions. In this year's conference, we hope to have contributed toward the creation of a long-lasting series of annual meetings in Málaga where researchers, companies, and even municipal authorities can find answers and advances for the final benefit of citizens and domestic economies. With a clear scientific focus, we also hold industrial demonstrations and promote bold thinking.

The topics of the papers in this volume include studies and tools to improve our knowledge on weather, road traffic, buildings, the behavior of citizens, logistics, frameworks to build and communicate in new services, sensors, simulations, and a wide spectrum of new urban technologies for our near future.

We thank Springer for helping with these proceedings, as well as the University of Málaga for their continuous support in the organization. We also thank the Spanish project moveON (TIN2014-57341-R) and the Spanish network of universities for smart cities CI-RTI (TIN2016-81766-REDT) for endorsing this activity, as well as all the enthusiasm of the members of the NEO research group in Málaga.

June 2017

Enrique Alba
Francisco Chicano
Gabriel Luque

Organization

Organizing Committee

General Chair
Enrique Alba   University of Málaga, Spain

Program Chair
Gabriel Luque   University of Málaga, Spain

Publication Chair
Francisco Chicano   University of Málaga, Spain

Local Organization
Jamal Toutouh   University of Málaga, Spain
Javier Ferrer   University of Málaga, Spain
Javier Arellano   University of Málaga, Spain
Daniel Stolfi   University of Málaga, Spain
Yesnier Bravo   University of Málaga, Spain
Zakaria Abd El Moiz Dahi   Université Constantine 2, Algeria
Andrés Carneo   University of Málaga, Spain

Publicity Chair
Christian Cintrano   University of Málaga, Spain

Program Committee

Enrique Alba   University of Málaga, Spain
Juan Antonio Alvarez-García   University of Seville, Spain
Rosa M. Arce   Universidad Politécnica de Madrid, Spain
Raquel Barco   University of Málaga, Spain
Pierfrancesco Bellini   University of Florence, DINFO, Italy
Juergen Branke   University of Warwick, UK
Nelio Cacho   Universidade Federal do Rio Grande do Norte, Brazil
Miriam Capretz   University of Western Ontario, Canada
Jose M. Carmona   Andata GmbH, Austria
J. Marcos Castro   University of Málaga, Spain
María Rosa Cervera Sarda   Universidad de Alcalá de Henares, Spain
Francisco Chicano   University of Málaga, Spain
Felipe Espinosa   University of Alcalá de Henares, Spain
Javier Faulin   Universidad Pública de Navarra, Spain

Contents

A Robust and Lightweight Protocol Over Long Range (LoRa) Technology
for Applications in Smart Cities 1
    Félix Sasián, Diego Gachet, Miguel Suffo, and Ricardo Therón

A Sustainable Bi-objective Approach for the Minimum Latency Problem. . . . 11
    Nancy A. Arellano-Arriaga, Ada M. Álvarez-Socarrás,
    and Iris A. Martínez-Salazar

Cities as Enterprises: A Comparison of Smart City Frameworks Based
on Enterprise Architecture Requirements 20
    Viviana Bastidas, Marija Bezbradica, and Markus Helfert

Comparative Study of Artificial Neural Network Models for Forecasting
the Indoor Temperature in Smart Buildings 29
    Sadi Alawadi, David Mera, Manuel Fernández-Delgado,
    and José A. Taboada

Considering Congestion Costs and Driver Behaviour into Route
Optimisation Algorithms in Smart Cities 39
    Pablo Alvarez, Iosu Lerga, Adrian Serrano, and Javier Faulin

Distributed Genetic Algorithms on Portable Devices for Smart Cities 51
    J.A. Morell and Enrique Alba

Existing Approaches to Smart Parking: An Overview 63
    Fernando Enríquez, Luis Miguel Soria, Juan Antonio Álvarez-García,
    Francisco Velasco, and Oscar Déniz

Impact of Protests in the Number of Smart Devices in Streets:
A New Approach to Analyze Protesters Behavior 75
    Antonio Fernández-Ares, Maria Garcia-Arenas,
    Pedro A. Castillo, and Juan J. Merelo

Logistics Support Approach for Drone Delivery Fleet 86
    Asma Troudi, Sid-Ali Addouche, Sofiene Dellagi,
    and Abderrahman El Mhamedi

Policy Recommendations Supporting Smart City Strategies:
Towards a New Methodological Tool 97
    Nils Walravens and Pieter Ballon

Predicting Car Park Occupancy Rates in Smart Cities 107
    Daniel H. Stolfi, Enrique Alba, and Xin Yao

Predicting Individual Trip Destinations with Artificial Potential Fields   118
  Alessandro Zonta, S.K. Smit, and Evert Haasdijk

Robust Bi-objective Shortest Path Problem in Real Road Networks   128
  Christian Cintrano, Francisco Chicano, and Enrique Alba

Smart Urban Mobility from Expert Stakeholders' Narratives   137
  Daniel Lopatnikov

Simulation Model of Traffic in Smart Cities for Decision-Making Support:
Case Study in Tudela (Navarre, Spain)   144
  Juan-Ignacio Latorre-Biel, Javier Faulin, Emilio Jiménez,
  and Angel A. Juan

Virtual Development of a Presence Sensor Network Using 3D Simulations . . .   154
  Rafael Pax, Marlon Cárdenas Bonett, Jorge J. Gómez-Sanz,
  and Juan Pavón

Author Index   165

A Robust and Lightweight Protocol Over Long Range (LoRa) Technology for Applications in Smart Cities

Félix Sasián[1], Diego Gachet[2($\boxtimes$)], Miguel Suffo[3], and Ricardo Therón[4]

[1] Desing3 SL, Recinto Interior Zona Franca Nave 1-A1, Cádiz, Spain
felix.sasian@desing3.com
[2] Universidad Europea de Madrid, 28670 Villaviciosa de Odón, Spain
diego.gachet@universidadeuropea.es
[3] Department of Mechanical Engineering and Industrial Design, Engineering School,
Universidad de Cádiz, Polígono Río San Pedro s/n, 11510 Puerto Real, Spain
miguel.suffo@uca.es
[4] Carbon Management - Solar Energy, Research & Development Center,
Saudi Aramco, Dhahran, Saudi Arabia
ricardo.theron.basta@icloud.com

Abstract. The implementation of Smart Cities requires technologies such as the Internet of Things (IoT). This paradigm aims to integrate and use in Internet electronic embedded devices like sensors, in order to share information and allowing interaction from anywhere. Standard communication protocols used in IoT for data transmission are based in general on FSK (Frequency Shift Keying) techniques that do not have the range and necessary performance to deal with this new scenery. With this situation in mind Semtech Technologies have developed a new modulation technique called LoRa (Long Range) based on spread spectrum that offer a new communication model for low power and long range networks, impossible to implement with other techniques. In this paper, we describe a new communication protocol implemented over the physical layer of LoRa providing effectiveness, robustness and low power consumption suitable for IoT applications in the context of smart cities. Results obtained related with the coverage analysis confirm the success of the selection of this technology and opens new horizons for developing useful applications in domains like smart cities, e-health, smart factory, remote monitoring and control, etc.

Keywords: Internet of things · Cloud computing · Wireless protocol · LoRa technology · Smart City · Sensors · Big data

1 Introduction

Traditional planning, acquisition and funding of current cities are not adequate to meet the requirements of intelligent cities [1]. The cities or territories need a new economic model, the adoption of more efficient energy practices and providing services that are more accessible to citizens. This new development model, however, requires a suitable environment for the adoption and effectively use of smart solutions. Therefore, the International Electrotechnical Commission and the International Telecommunication

© Springer International Publishing AG 2017
E. Alba et al. (Eds.): Smart-CT 2017, LNCS 10268, pp. 1–10, 2017.
DOI: 10.1007/978-3-319-59513-9_1

Union, with the participation of European Telecommunications Standards Institute, coordinate these objectives, shared by all those involved in projects for the deployment of smart networks.

The definition of a Smart City remains controversial. While the overall concept is generally perceived as an ideal fusion of sustainability with advanced technology, there is some confusion regarding the place that the term "intelligent" deals between "sustainable", "strong" and "resilient" [2]. An easy to use transport system, fluid road networks, energy efficiency, clean air pollution, water and waste management, the respect to the environment and the adoption of effective measures to protect the safety and health of citizens, etc. are some of indicators to take into account when we are talking about Smart Cities among others important drivers as for example Technologies, Social structures, Cultural drivers, Economic drivers, Governance, etc. Authors in [3], provide a first operational definition of Smart Cities, grouping 74 indicators in six categories. The European Union (EU) is investing a great amount of resources in developing a strategy for a "smart" urban growth for its metropolitan regions and, consequently, has developed a series of actions included in the Digital Agenda for Europe. In fact, the Horizon 2020 is the biggest EU Research and Innovation programme of the European Commission giving the opportunity to private and public actors, as well as the researchers, to propose projects related to the Smart Cities.

There are some successful examples of technologies and programs of Smart Cities through urban innovation labs, which leverage public knowledge to identify and solve local problems using available open data repositories. These examples should serve as a standardized model for organizing and manage future changes in a city. The World Council on City Data Open Data Portal [4] allows us to explore, track, monitor and compare several cities on up to 100 service performance and quality of life indicators. Recently the ISO/WG 268 work group published the ISO 37120 and ISO 37101 standards "Sustainable development and resilience of communities. Indicators for city services and quality of life", for helping communities to develop and implement management systems for improve their actions in sustainable development.

According to [5], Spain is the first European country in number of projects related with Smart Cities, as the first tourism destination in the world, the second in tourism expenditure and third by number of international tourists, 11% of Spanish GDP (Gross Domestic Product) is due to the tourism industry. Therefore, since 2012, there is an interest in political authorities for launching, with the help of ICT (Information and Communication Technologies), innovation projects related to Smart Cities for creating differentiated services and highly competitive smart destinations. From this year, there is a Technical Standards Committee AEN/CTN 178 "Smart Cities" within the Spanish Association for Standardization and Certification that has incorporated the aforementioned ISO 37120 standards to national regulations, for enabling application compatibility and allowing operation and development of citizen services by different entities not related with developers of communications platforms. The purpose of this paper is to describe a technological approach related with data communications to be used in the Smart Cities context.

2 Related Work

There is a close relationship between the Smart Cities concept and the IoT, specially related to the wireless communication technology [6], if we are considering the necessity of sensing the city, it is clear that a huge set of heterogeneous sensors need to communicate among them and with others devices. The problem, is that at present time there is a range of technologies to be used but there is no a defined standard, in this sense a further relevant IETF (Internet Engineering Task Force) Working Group named Routing Over Low power and Lossy networks (ROLL) has recently write the RPL (IPv6 Routing Protocol for Low-Power and Lossy Networks) draft. This will be the basis for routing data over low power and lossy networks including 6LoWPAN, which still needs many contributions to reach a full solution and to reduce the uncertainty in their implementation.

A good comparison of communication technologies used in the context of Smart Cities can be found in [7]. Using as example the intelligent management of solid urban waste, some interesting data can be extracted from that analysis, in the sense that there are some important variables to be considered in the context of a Smart City, for example the scope, consumption, range and specially the ownership of the communication networks. This last aspect can often be decisive since, in most cases, these technologies are protected by patents and ultimately involve the payment of royalties for the development of new deployments. The graph in Fig. 1 shows the currently existing market niche between short and long-range communications and low rate data packet.

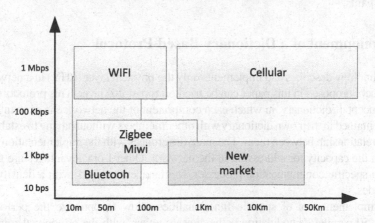

Fig. 1. Long range and short-range connectivity

The current panorama of existing technological solutions for application in wireless sensor networks is wide. It makes possible to perform a multicriteria selection that suits to the "smart" ecosystem needs. Due to requirements of low battery consumption, embedded sensors are preferred in the nodes themselves, and with the capability to communicate with very low data rate. This is especially important in applications like e-health or tele-monitoring [8]. For example GPS/GPRS networks appear having a great bandwidth, but with a high installation cost, high-energy consumption and, the most important aspect, the cost of connection. SigFox [9], is also a proprietary ultra-narrow band technology, it has very long range and low power consumption, but with a very limited capacity to transmit information, as it is focused on sending small messages. In addition, as it is a relatively new service model in a market dominated by other technologies, there are uncertainty in the survival and expansion of the network. The previous-generation technologies such as ZigBee, Bluetooth or Miwi networks are focused on short-range (below 100 m) and do not have a high-energy efficiency.

Finally, Lora WAN [10] is a specification for LPWAN (Low Power WAN) network that provides bi-directional communication and, is focused on battery-powered devices. It is supported by the LoRa Alliance, a non-profit organization made up of several companies that collaborate in the development of the common protocol. LoRa WAN is a good choice with excellent range and energy consumption, and it offers free access to source code. However, its use, involves implementing the complete specification, regardless of the needs that a particular application requires. Moreover, at present, this specification is not yet established as a reference in the market and other competitors develop and offer alternative protocols to LoRa over their physical layer that is the case of Lab Link with the so-called Symphony Link.

With LoRa technology in mind, this paper proposes a new lightweight, flexible and secure protocol, called DBP (Dictionary Based Protocol). This protocol is developed over the physical layer of LoRa and it is based on the concept of dictionary. It has the advantage of allowing the exchange of information in a simple and efficient way to nodes, from small local networks with few components, to large networks of thousands of participants.

3   Development of a Dictionary Based Protocol

LoRa technology describes and implements only the physical layer (PHY) in a network, the new protocol proposed in this paper can be used on top of this layer. This protocol is based on the concept of dictionary, in which each component of the network can exchange information contained in their own dictionary with other members without having to establish any previous relationship between them. The main differences with the implementation on FSK [11] lie in the capacity for addressing in the network. Other Lora advantages are the existence of a specific communication processor, to manage a network with a density of 10–50000 nodes.

The implementation of specific functionalities allow to optimize the performance, including 50 parallel demodulation paths, that, together, with the orthogonal modulation scheme permits a level of coexistence in the network unattainable with solutions based on FSK or OOK (On-Off Keying). The implementation of DBP (Dictionary Based Protocol) over LoRa transceivers allows it to be used in situations where a long range (even over 10 km) and networks with large numbers of nodes are needed.

Given the protocol's characteristics, any member of the network can send and receive information directly from any other member, so it is possible to use point-to-point or point-to-multipoint communication models, although DBP is focused on star networks with one or more gateways. The applicable topologies for a DBP protocol are shown in Fig. 2.

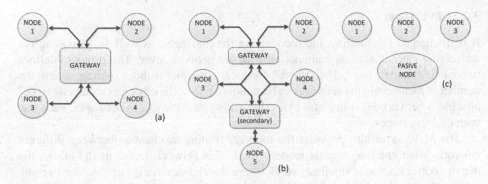

Fig. 2. Topologies for DBP protocol: (a) traditional scheme with a coordinator, (b) multi - coordinator system, (c) producer-consumer scheme.

3.1 Physical Layer

The physical layer is used by the transceiver to synchronize the communication and ensure its viability. It is based on the Semtech LoRa chipset [12]. Messages are in "explicit mode", in which the length of the data is not always fixed and is not known in advance. The frame of a LoRa package begins with the preamble, followed by the sync byte and a header that includes the length of the message and a CRC. The detail of a complete DBP packet is shown in Fig. 3.

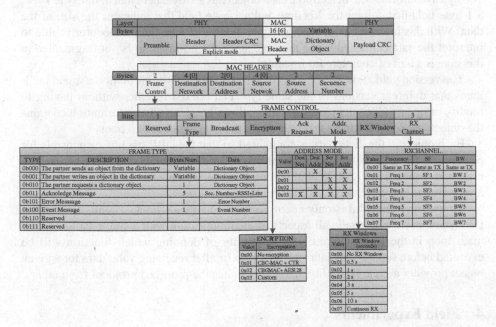

Fig. 3. DBP protocol packet detail.

3.2   MAC Layer

It is in charge of delimiting and recognizing the messages, as well as, providing the addressing mechanisms, in conjunction with the network layer. The protocol defines two routing modes, one called "short", for local networks without routing, where the number of members in the network can be at most (2^{16}-2). The other mode named "large" also has a network identifier of 4 bytes that allows to deploy (2^{32}) networks, each one with (2^{16}-2) devices.

The MAC layer also provides the message routing mechanism between different networks when the long routing mode is used. This network access model allows the use of shorter frames in small networks where the end devices do not require to route information outside their own network, or where a network identifier is not even necessary.

To make use of the LoRa capabilities, it is possible to establish a receive window after each sending, and even receiving a configuration of parameters different from those used in the transmission. DBP allows signing and encrypting messages [11]. On DBP over LoRa, it is possible to use XTEA, AES or another proprietary algorithm to perform the encryption.

3.3   The Dictionary

The dictionary is an ordered set of objects, addressed by an 8-bit index and no more than 100 bytes of information. In addition, each object has a 1-byte field that defines the Type, a 1-byte field that defines the Version, and a 1-byte field that indicates the size of the data. With this information of type, version, index and size, the interlocutor is able to interpret the data contained in the message since the information is not fragmented (a message is a unit of complete information).

The version field allows developers to perform revisions of his dictionary, and facilitates that different versions of the same type can coexist together without having to modify or update all equipment. The index allows knowing what information contains the object; we could compare this index with the page number of a book.

Each object in the dictionary is configured so that the DBP stack is responsible for performing all tasks related to communication. This configuration is recorded at each node and is shown in Fig. 4. With this configuration capability in each object, we have a very simple and extremely powerful, and flexible system.

To use the protocol it is only necessary to define the dictionary configuration, the DBP stack itself will handle all aspects of communication with a single call from the main loop in the user program. The possibility of defining which function will be executed before sending the data after a request or after receiving valid data for a given object provides a great flexibility that differentiates the proposed protocol from others.

4   Field Experiments

Using the LoRa technology explained before, a testing platform formed by hardware and software components was developed. Such components are the following:

Dictionary Objects Config Options	
Property	Description
Index	Index of the object.(Should be unique and not necessarily consecutive)
Numbyte	Number of bytes of the object (only data)
RWMode	Accessmode.(RW,ReadOnly,WriteOnly)
CryptoOnly	Only encrypted access allowed
UnicastOnly	Only unicast access allowed (broadcast forbidden)
*Data	Pointer to data area
*PrevFunction	Pointer to function to be executed before sending data from the object
*PostFunction	Pointer to function to be executed after receive valid data for the object

Fig. 4. Configuration data for dictionary

Fig. 5. Gateway based on SX1301.

- A base station that performs a gateway functions between the wireless devices and the application server. The station has been installed over a communications tower located 40 m high by using a 3G modem.
- Several end devices or nodes, which are powered by batteries and implemented through a small microcontroller. Each of those nodes is equipped with a LoRa technology modem.
- A terminal application programmed in C# language running on Windows platform was developed for collecting, representing and analyzing the received information.

As shown in Fig. 5, the base station is formed by the following components:

- IC880A-USB (SX1301 + 2x SX1557) Card. Figure 5(a)
- A Raspberry PI with control application based on Linux. Figure 5(b) illustrates this component located inside a cabinet designed to perform the tests, whereas Fig. 5(c) and (d) show the cabinet located on top of the 40 m high tower.
- An ISM band base antenna with a gain of 5 dBi. Figure 5(b) represents this component installed inside the testing cabinet.
- A 3G Teltronika RUT500 Router, which is also shown in Fig. 5(b).

The SX1301 management software is implemented on Linux (although we also developed a Windows version). This software is based on the public information given by the manufacturer and a Raspberry PI card connected via a USB port to facilitate a rapid implementation of the software carries out its execution.

Moreover, in order to have a more stable application, the use of a 32-bit microcontroller connected through a SPI port is recommended. The end nodes are based on the PIC24f32KA302 microcontroller, which is specifically designed for ultra-low power consumption, being powered by two AAA size batteries.

5   Results and Discussion

Up to now, it is not possible to have reliable data on the real coverage of a LoRa LPWAN based network. However, it is possible to find in the literature some tests based on point-to-point fixed geographical locations used for this purpose. It is clear, therefore, that the development of the proposed hardware prototypes and the DBP protocol is an important contribution to this field.

The prototype has been used to perform a full analysis of coverage both, in urban and rural environments. As noted above, to carry out the tests, the base station was installed on a 40 m telecommunications tower located in the city of Puerto Real, with a total height over sea level of 94 m.

For final nodes, we have installed two transmitters on a vehicle, during a journey; a continuous sampling of more than twelve thousand messages was done using Spreading Factor from SF7 to SF12. The data transmitted is the current GPS position, with this configuration was possible to know the RSSI (Received Signal Strength Indicator) and LSNR (Low Signal to Noise Ratio) measured for each of traveled geographical places. If we do a superposition of all information on a Google satellite map, it is easy to observe a clear view about the capacity of the LoRa link for any point inside the studied geographical area. Our group has developed a C# application for data analysis using GMap.net API.

The values collected using the proposed prototype confirms the initial data given by the manufacturer; these results were based on "line of sight" applications on a 10 km radius and in "No line of sight" with 3 km radius. Figure 6 shows a geographical map with real results of the range reached for the total considered sampling points in the test done in rural and urban zones. It is important to note the value of RSSI = −100 dBM on the town of Medina Sidonia 22 km away from the base station.

TIME: 2015-12-08 T09:48
LAT:36,4688350
LON: -5.9294200
SPEED: 5,6 Km/h
RSSI: -98 dBm
LSNR: 72dB
SF&BW: SF11BW125
Distance: 22,53 Km

Fig. 6. Map of real coverage obtained as result of test in urban and rural areas around Cádiz

6 Conclusions

This paper presents a new protocol for wireless network implemented on LoRa technology, the main characteristic is that is based on the concept of a dictionary, in other words this imply that the nodes exchange information like object previously defined in their own dictionary. This concept simplifies de design and implementation of small and big scale wireless networks from a few to thousand nodes.

The protocol also implements the necessary mechanism for signing and encryption of communications. The results obtained in the coverage tests show that the LoRa based protocol is a viable alternative to be used in the new Smart Cities market, e-health applications and remote monitoring, it is clear that the protocol could be used in applications like street lighting control, water and waste control, or like smart factory, etc.

The LoRa network can be deployed with an initial low cost because a gateway can control an urban or rural area up to 10 km of radius, if it is installed in an adequate geographical area, it is important the availability of a isocoberture map for identify shadow zones. The features mentioned for the new protocol combined with the low cost, scalability, flexibility and public availability of LoRa specifications make the proposed prototype a valid solution for new developments of wireless-sensors network applications.

Acknowledgments. This work is still being developed through the support of Desing3 S.L Company. Some aspects of this project were studied through funds granted by the Spanish Ministry of Economy and Competitiveness under project iPHealth (TIN-2013-47153-C3-1).

References

1. Letaifa, B.S.: How to strategize Smart Cities: revealing the SMART model. J. Bus. Res. **68**(7), 1414–1419 (2015)
2. UN-Habitat. The City Resilience Profiling Programme, CRPP (2012). http://unhabitat.org/urban-initiatives/initiatives-programmes/city-resilience-profiling-programme, Accessed 20 Nov 2016
3. Giffinger, R.: Smart Cities. Ranking of European medium-sized cities. October, vol. 16, pp. 13–18 (2007)
4. WCCD ISO 37120. World Council City Data (2015). http://www.dataforcities.org/
5. UNWTO. UNWTO Tourism Highlights, 2015 edn. (2015). http://mkt.unwto.org/publication/unwto-tourism-highlights-2015-edition, Accessed 10 Dec 2016
6. Atzori, L., Lera, A., Morabito, G.: The Internet of Things: a survey. Comput. Netw. **54**(15), 2787–2805 (2010). Elsevier
7. Hannan, M.A., Al Mamun, M.A., Hussain, A., Basri, H., Begum, R.A.: A review on technologies and their usage in solid waste monitoring and management systems: issues and challenges. Waste Manage. **43**, 509–523 (2015)
8. Páez, D.G., de Buenaga Rodríguez, M., Sánz, E.P., Villalba, M.T., Gil, R.M.: Big data processing using wearable devices for wellbeing and healthy activities promotion. In: Cleland, I., Guerrero, L., Bravo, J. (eds.) IWAAL 2015. LNCS, vol. 9455, pp. 196–205. Springer, Cham (2015). doi:10.1007/978-3-319-26410-3_19
9. Margelis, G., Piechocki, R., Kaleshi, D., Thomas, P.: Low throughput networks for the IoT: lessons Learned from industrial implementations. In: Proceedings 2015 IEEE World Forum on Internet of Things (WF-IoT) (2015)
10. Semtech. LoRa Technology (2015). http://www.semtech.com/wireless-rf/lora.html, Accessed 10 Nov 2016
11. Sasián, F., Theron, R., Gachet, D.: Protocolo para comunicación inalámbrica en instalaciones de energías renovables. Revista Iberoamericana de Automática e Informática Industrial RIAI **13**(3), 310–321 (2016)
12. Semtech, Sx1272/73. Datasheet (2015). http://www.semtech.com/images/datasheet/sx1272.pdf, Accessed 12 May 2015

A Sustainable Bi-objective Approach for the Minimum Latency Problem

Nancy A. Arellano-Arriaga[1,2(✉)], Ada M. Álvarez-Socarrás[1],
and Iris A. Martínez-Salazar[1]

[1] Facultad de Ingeniería Mecánica y Eléctrica,
Universidad Autónoma de Nuevo León,
San Nicolás de los Garza, Nuevo León, Mexico
{nancy.arellanoarg,iris.martinezsalaz}@uanl.edu.mx,
nancy.arellano@uma.es, ada.alvarezs@uanl.mx
[2] Departamento de Economía Aplicada (Matemáticas),
Universidad de Málaga, El Ejido, Málaga, Spain

Abstract. Nowadays, sustainability is a major factor to consider in the decision-making process. Specifically, for companies trying to stay competitive and having some advantage in the market it is a vital issue. In this study, we introduce a multi objective problem which aims to minimize distance and latency of a route with enough capacity to serve a set of clients. We assume that a vehicle leaves an established depot, visits all clients and returns to the depot before the end of the workday. With this bi-objective problem, we aim to improve the sustainability of the company by improving their economic and environmental contribution, through the minimization of the traveled distance of the vehicle along with the improvement of their social service by the minimization of the total waiting time of the customers. We call this problem Minimum Latency-Distance Problem (MLDP) and in this paper, we introduce a mathematical formulation which describes it.

Keywords: Combinatorial optimization · Mathematical formulation · Multiple objective programming · Multi objective optimization · Multi objective problem · Latency · Distance · Multi objective routing problem

1 Introduction

Humanity developed the notion of sustainability by the awareness of the damage caused by living in big cities, the waste of natural resources among other environmental issues [1–4]. Sustainability is nowadays a global concept which rests on three important pillars: the environmental, the economic and the social. The first pillar refers to climate protection alongside the protection of the natural resources and the diversity of Earth's flora and fauna [5,6]. The second pillar refers to business and industries by focusing in regulating them with an agreement of responsibility towards the environment and the community. This

© Springer International Publishing AG 2017
E. Alba et al. (Eds.): Smart-CT 2017, LNCS 10268, pp. 11–19, 2017.
DOI: 10.1007/978-3-319-59513-9_2

agreement deals with the reduction of the waste and the environmental pollution the companies generate as well as to improve the contribution of all social improvements created by the companies [7–9]. The last pillar of sustainability refers to the equality among humanity by referring to the importance of all human beings, their integrity, and in general, the satisfaction of their needs within ecological constraints [10, 11]. Summarizing, sustainability attempts to protect the environment and the human beings integrity meanwhile respecting the structure and the resources of the Earth.

It is known that in general, deterministic relationships between the sustainability pillars are inadequate and lead to make several trade-offs [12, 13]. Dobson [14, 15] states that environmental and social sustainability are not compatible, this means that to gain in one pillar we have to lose in another, and concludes that deterministic decisions are only served in an incomplete way. Therefore, it is important to introduce new approaches to already known problems which attempt to aim several objectives focusing on the improvement of all pillars at the same time, targeting to the generation of more sustainable conditions for everyone. Multi-objective optimization specializes in taking decisions based on several objectives [16] and to choose a suitable trade-off for all involved stakeholders in the decision making.

In this paper, we present a bi-objective problem which arises in the context of logistic activities of distribution-and-service companies which specialize in customer service by attending requests of product delivery and/or maintenance services. We propose a bi-objective approach which attempts to obtain a trade-off that benefits all involved pillars of sustainability: environmental, economic and social, by simultaneously minimizing distance and latency of a route under certain assumptions.

The proposed problem is a direct application of two well-known routing problems: the minimum latency problem [17–19], and the traveling salesman problem [20]. The first one focuses on minimizing the total waiting time of a set of clients in a route, regardless the traveled distance of the vehicle. Meanwhile, the second problem deals with minimizing the total traveled distance of the vehicle, regardless the waiting time of the clients. By combining both objectives, we propose a client-centered approach which considers equally important the company resources and the quality of the service, or the social resources, in the decision making. The contribution to the environmental pillar of sustainability, goes along the minimization of the traveled distance of the vehicle by assuming the correspondence of less moving distance, fewer emissions the vehicle generates.

The main contribution of this paper is the development of a mixed integer formulation to represent this bi-objective problem. In Sect. 2, the description of the proposed problem is presented as well as its mathematical formulation. In Sect. 3, the computational experimentation is presented. And lastly, the conclusions are shown in Sect. 4.

2   Minimum Latency-Distance Problem

Let us consider a company in charge of attending a set of customers which demands a service, for example delivery or a maintenance issue; this company has an agent or a vehicle which is responsible of attending all the requests. This agent searches for a route that leaves from a known depot, visits all clients and returns to the depot, minimizing the travel distance of the vehicle, as well as the total clients' waiting time, or total latency of the route. All service times he may take to fullfill the request in each client and all travel times among clients are known, this is to say, every service and travel time are known before hand.

This bi-objective problem will be referenced from this moment on as the *Minimum Latency-Distance Problem* (MLDP). Due to the differences between the objectives involved in the MLDP and the fact that service and travel times are non-zero in real applications, in literature, it is known that both, distance and latency objectives are not calculated with the same metric [18,21]. Therefore, we assume several asseverations to formulate a model to describe the MLDP:

- For simplicity, we assume that traveling time is linearly proportional to traveled distance. Hence, the larger the traveled distance, larger the traveling time required to reach the client. This is a major assumption and in terms of applications, this is not always a true statement. We will assume all clients are reachable in a constant amount of time and the vehicle in charge of visit them will travel at the same speed all the time.
- One route is enough to serve all clients. Hence, this vehicle has infinite capacity.
- We assume the proportionality of the environmental savings to the traveled-distance savings. This is to say, less traveled distance, less emissions are generated by the vehicle.
- The environmental savings are not responsibility of the company in charge of providing the service to the clients. Environmental savings are a direct consequence of the minimization of the traveled distance, as previously stated.

Please note that this bi-objective approach has not been studied in the operations research area, and therefore it is a base model to consider the non-linearity of the objectives as future research. Previous studies [22], showed that it is not convenient to use a formulation designed for dealing with the objective of minimizing distance to handle an objective of minimizing the total latency of a route. Therefore, for the modelling of the MLDP, we take advantage of a model developed for the objective of latency. The issues related to the distance objective are later incorporated. Reported in literature, there are several formulations for the minimum latency problem. Some examples are the formulation presented in Méndez Díaz et al. [23], Gouveia and Voß [24], Picard and Queyranne [25], the implementation of Picard's formulation by Sarubbi et al. [26], and lately, Angel-Bello et al. [21]. However, [21] outbid them all by proposing two mathematical models that showed a better performance than all existing formulations. Therefore, our formulation takes as reference one of those models, called by Angel-Bello et al. [21] as "Model A".

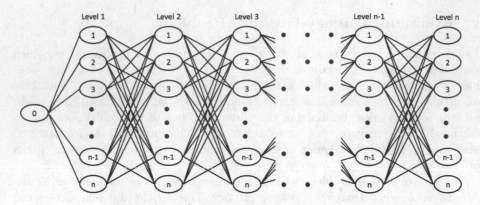

Fig. 1. Multi-level network used to formulate the MLDP.

Consider a directed and complete graph $G = (V, A)$. Let A be the arc set and $V = \{\{0\} \cup I\}$ the vertex set, where 0 is defined as the depot and the subset $I = \{1, 2, \ldots, n\}$ are the clients meant to be visited. Each client $i \in I$ has a service time s_i associated and each arc $(i, j) \in A$ has associated a travel time from client i to j, defined as t_{ij}. The connections among the depot and all the clients are also considered. For the formulation, a cost matrix $C = c_{ij}$ is built, considering:

$$c_{ij} = \begin{cases} t_{0j}, & \text{for } i = 0; j = 1, 2, \ldots, n; \\ s_i + t_{ij}, & \text{for } i = 1, 2, \ldots, n; j = 0, 1, \ldots, n; j \neq i. \end{cases}$$

This formulation is based on a multi-level network (see Fig. 1) with $n + 1$ levels inspired by one previously proposed by Picard and Queyranne in 1978, which was designed for a time-dependent travelling salesman problem [25].

In this multi-level network level 0 represents the depot, while levels $1, 2, \ldots, n$ are n copies of the set of clients. Note that an MLDP solution is a permutation of the n clients with node 0 in the first position of the permutation. Consequently, any solution can be represented on this multilevel network having a single node active in each level, with no repetitions between levels. These facts entail to define the decision variables x_i^k and y_{ij}^k as

$$x_i^k = \begin{cases} 1, & \text{If node } i \text{ is selected on level } k \text{ on the network,} \\ 0, & \text{otherwise.} \end{cases}$$

$$y_{ij}^k = \begin{cases} 1, & \text{If node } j \text{ on level } k + 1 \text{ follows node } i \text{ on level } k, \\ 0, & \text{otherwise.} \end{cases}$$

Therefore, MLDP is then formulated as:

$$min \ F_1 = n \sum_{i=1}^{n} c_{0i} x_i^1 + \sum_{k=1}^{n-1} \sum_{i=1}^{n} \sum_{j=1}^{n} (n-k) c_{ij} y_{ij}^k \tag{1}$$

$$min \ F_2 = \sum_{i=1}^{n} c_{0i} x_i^1 + \sum_{k=1}^{n-1} \sum_{i=1}^{n} \sum_{j=1}^{n} c_{ij} y_{ij}^k + \sum_{i=1}^{n} c_{i0} x_i^n \tag{2}$$

s.a

$$\sum_{k=1}^{n} x_i^k = 1, \ i = 1, 2, \ldots, n \tag{3}$$

$$\sum_{i=1}^{n} x_i^k = 1, \ k = 1, 2, \ldots, n \tag{4}$$

$$\sum_{\substack{j=1 \\ j \neq i}}^{n} y_{ij}^k = x_i^k, \ i = 1, 2, \ldots, n, \ k = 1, 2, \ldots, n-1 \tag{5}$$

$$\sum_{\substack{j=1 \\ j \neq i}}^{n} y_{ji}^k = x_i^{k+1}, \ i = 1, 2, \ldots, n, \ k = 1, 2, \ldots, n-1 \tag{6}$$

$$x_i^k \in \{0, 1\}, \ i = 1, 2, \ldots, n, \ k = 1, 2, \ldots, n \tag{7}$$

$$y_{ij}^k \geq 0, \ i = 1, 2, \ldots, n, \ j = 1, 2, \ldots, n, j \neq i, \ k = 1, 2, \ldots, n \tag{8}$$

The minimization of the total waiting time of the clients is defined by Eq. (1). This Equation is a mathematical representation of the social interest in this proposed problem. In Eq. (2) lays the minimization of the total traveled distance and therefore, the minimization of the cost this route represents to the company. By assuming that minimizing traveled distance is proportional to minimizing emissions, the minimization of the emissions generated by the vehicle goes along this equation and therefore, this Equation is the mathematical representation of the economic and environmental interests in this proposed problem. Note that in Eq. (1) the arc denoting the return of the vehicle to the depot is not considered, meanwhile this returning arc has to be considered on Eq. (2). The reason of this is we do not consider the returning of the vehicle to the depot as client's waiting time.

Equation (3), guarantees that each node occupies a single position in any feasible solution. Equation (4), guarantees that each position is occupied by no more than one node in any feasible solution. Equation (5), ensures that only one arc leaves from position k, exactly from the node taking that position and Eq. (6) imposes that only one arc at a time can arrive to position $k+1$, exactly to the node occupying that position. The nature of Eqs. (3) to (6), lays in the fact that all possible solutions for the MLDP can be represented on the multi-level network previously defined.

Lastly, Eqs. (7) and (8) correspond to the nature of the variables. Note that variables y_{ij}^k are previously defined as binary but they are handled as continuous variables in the model. This is achieved by Eqs. (5) and (6), which guarantee that only one arc leaves from position k and arrives exactly to the node occupying the position $k+1$. These pair of constraints assures y_{ij}^k takes a value of one or a value of zero. This formulation consists of n^2 binary variables, $n^3 - 2n^2 + n$ real variables and $2n^2$ constraints.

3   Computational Experimentation

Because both single-objective problems involved are NP-hard [27,28], it is known before hand that to obtain a true Pareto front, it could take an exponential-crescent time according to the size of the instance. The study of the complexity for this proposed problem is left for future research. To test the mathematical formulation, we implemented the *Augmecon2* procedure, proposed by Mavrotas et al. [29].

The Augmecon2 procedure, in general, fixes one objective and attempts to solve the multi-objective formulation as a single objective one, considering all the original constraints plus the rest of the objectives as constraints as well. In our particular implementation of Augmecon2, we defined the distance objective as the main objective the one to pursue. The algorithm requires the definition of a step, in our particular implementation this step is defined as the difference between the optimal solution of the mono-objective latency problem and the optimal distance solution, evaluated on the function of latency, divided by 10 as it was proposed by Mavrotas et al. [29].

We tested our implementation with the same group of instances used on the work of Angel-Bello et al. [21]. These instances are called *GTRP* and were obtained by randomly generated points with real coordinates from a uniform distribution between 0 and 100. The travel times defined in these instances were taken as Euclidean distances and rounded down to integers. Service times s_i were considered as zero (GTRP-S_0) and non-zero, defined in between the interval $[0, (t_{max} - t_{min})/2]$ (GTRP-S_1). In these intervals, $t_{max} = max\{t_{ij}\}$ and $t_{min} = min\{t_{ij}\}$. We maintained the original classification of the instances by the type of service times they included (GTRP-S_0 and GTRP-S_1). For each type, each set of instances has several sizes: 10, 15, 20, 25, 30, 35, 40, 60, 80 and 100 clients, each size with 25 instances. All selected instances were solved in a Xenon ®Intel ®CPU E3-1245 v3 @ 3.40 GHz, with 16.0 GB of RAM and the MILP solver selected was ILOG CPLEX C++ Concert Technology.

The selected limit time on each instance was 10,800 s, three hours, CPU Times taken by this algorithm can be seen on Table 1. Note that for both service times we could only reach the true Pareto Front for instances of 25 clients, for larger instances we reached the limit time before finding the exact frontier. In Fig. 2, an example of a 25-client instance exact Frontier is depicted. Note that to select a single solution in a Pareto front, it is advisable to review all existing methodologies and select the most appropriate [30].

Table 1. Average CPU Times obtained using Augmecon2 procedure.

Instance	Size	CPU Time (s)
GTRP-S_0	10	15.340
	15	113.678
	20	1274.710
	25	4263.970
GTRP-S_1	10	18.610
	15	122.830
	20	1587.842
	25	10195.500

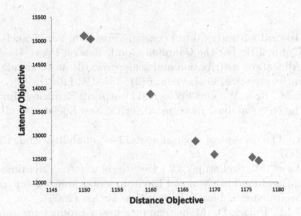

Fig. 2. Points of the exact Pareto front of a GTRP-S_0 instance of size 25, found with the described implementation of the Augmecon2 procedure.

4 Conclusions

In this paper, we introduced a bi-objective problem that seeks to deal with sustainability issues within a routing problem. We propose to minimize the traveled distance and the total latency of a route, to minimize the environmental impact, to improve the economy of a business and to improve the social interaction among the company and the clients, by the integration of the client in the decision-making process. This approach is useful, not only to reduce the cost that the traveling represents to the company, but to improve customer service by minimizing the total waiting time of all the clients. By bringing together the economic and social aspects of this problem, the company can obtain bigger benefit in general.

We presented a mathematical formulation for the MLDP problem and solved it to obtain exact Pareto fronts. We were able to solve instances up to 25 clients with both service times, zero and non-zero, before reaching limit time.

In future research, it is important to consider all involved times stochastic. It would be advisable to introduce a better way to consider the emissions this vehicle generates and to redefine the assumptions made about the linearity between traveled time and traveled distance. The introduction of several vehicles to visit all clients, as well as the introduction of capacity in the vehicles are other issues to add.

Acknowledgements. The first author would like to thank CONACYT, the Mexican National Council for Science and Technology, which supports her studies. This research did not receive any specific grant from funding agencies in the public, commercial, or not-for-profit sectors.

References

1. Daly, H.E.: Toward a Steady State Economy. Freeman, San Francisco (1973)
2. Daly, H.E., Cobb, J.B.: For the Common Good. Beacon Press, Boston (1989)
3. Daly, H.E.: Allocation, distribution and scale: towards an economics which is efficient, just and sustainable. Ecol. Econ. **6**(3), 185–193 (1992)
4. Wackernagel, M., Rees, W.: Our Ecological Footprint: Reducing Human Impact on the Earth. The New Catalyst. Bioregional Series. New Society Publishers, Gabriola Island (1998)
5. Goodland, R.: The concept of environmental sustainability. Ann. Rev. Ecol. Syst. **26**, 1–24 (1995)
6. Esty, D.C., Levy, M., Srebotnjak, T., De Sherbinin, A.: Environmental Sustainability Index: Benchmarking National Environmental Stewardship, pp. 47–60. Yale Center for Environmental Law & Policy, New Haven (2005)
7. Dangelico, R.M., Pujari, D.: Mainstreaming green product innovation: why and how companies integrate environmental sustainability. J. Bus. Eth. **95**(3), 471–486 (2010)
8. Baral, N., Pokharel, M.P.: How sustainability is reflected in the S&P 500 companies strategic documents. Organ. Environ. (April 24, 2016). doi:10.1177/1086026616645381
9. Schaltegger, S., Hansen, E.G., Ldeke-Freund, F.: Business models for sustainability: origins, present research, and future avenues. Organ. Environ. **29**(1), 3–10 (2016)
10. Polèse, M., Stren, R.: The Social Sustainability of Cities: Diversity and the Management of Change. University of Toronto Press, Toronto (2000)
11. Littig, B., Griessler, E.: Social sustainability: a catchword between political pragmatism and social theory. Int. J. Sustain. Dev. **8**(1), 65–79 (2005)
12. Goodland, R.: Sustainability: Human, Social, Economic and Environmental. Encyclopedia of Global Environmental Change. Wiley, London (2002)
13. Costanza, R., Graumlich, L., Steffen, W., Crumley, C., Dearing, J., Hibbard, K., Leemans, R., Redman, C., Schimel, D.: Sustainability or collapse: what can we learn from integrating the history of humans and the rest of nature? Ambio **36**(7), 522–527 (2007)
14. Dobson, A.: Justice and the Environment: Conceptions of Environmental Sustainability and Theories of Distributive Justice. Clarendon Press, Oxford (1998)
15. Dobson, A.: Fairness and Futurity: Essays on Environmental Sustainability and Social Justice. Oxford University Press, Oxford (1999)
16. Ehrgott, M.: Multicriteria Optimization. Springer, Heidelberg (2005)

17. Lucena, A.: Time-dependent traveling salesman problem-the deliveryman case. Networks **20**(6), 753–763 (1990)
18. Blum, A., Chalasani, P., Coppersmith, D., Pulleyblank, B., Raghavan, P., Sudan, M.: The minimum latency problem. In: Proceedings of the Twenty-Sixth Annual ACM Symposium on Theory of Computing (STOC 1994), pp. 163–171. ACM, New York (1994)
19. Chaudhuri, K., Godfrey, B., Rao, S., Talwar, K.: Paths, trees, and minimum latency tours. In: Proceedings of the 44th Annual IEEE Symposium on Foundations of Computer Science, pp. 36–45. IEEE (2003)
20. Laporte, G.: The traveling salesman problem: an overview of exact and approximate algorithms. Eur. J. Oper. Res. **59**(2), 231–247 (1992)
21. Angel Bello, F., Álvarez, A., García, I.: Two improved formulations for the minimum latency problem. Appl. Math. Model. **37**(4), 2257–2266 (2013)
22. Angel-Bello, F., Martinez-Salazar, I., Alvarez, A.: Minimizing waiting times in a route design problem with multiple use of a single vehicle. In: INOC (2013)
23. Méndez-Díaz, I., Zabala, P., Lucena, A.: A new formulation for the traveling deliveryman problem. Discrete Appl. Math. **156**(17), 3223–3237 (2008)
24. Gouveia, L., Voß, S.: A classification of formulations for the (time-dependent) traveling salesman problem. Eur. J. Oper. Res. **83**(1), 69–82 (1995)
25. Picard, J.C., Queyranne, M.: The time-dependent traveling salesman problem and its application to the tardiness problem in one-machine scheduling. Oper. Res. **26**(1), 86–110 (1978)
26. Sarubbi, J., Luna, H., Miranda, G.: Minimum latency problem as a shortest path problem with side constraints (2008)
27. Afrati, F., Cosmadakis, S., Papadimitriou, C.H., Papageorgiou, G., Papakostantinou, N.: The complexity of the travelling repairman problem. RAIRO Theor. Inf. Appl. **20**(1), 79–87 (1986)
28. Papadimitriou, C.M.: Computational Complexity. Addison-Wesley, Massachusetts (1994)
29. Mavrotas, G., Florios, K.: An improved version of the augmented -constraint method (augmecon2) for finding the exact pareto set in multi-objective integer programming problems. Appl. Math. Comput. **219**(18), 9652–9669 (2013)
30. Ferreira, J.C., Fonseca, C.M., Gaspar-Cunha, A.: Methodology to select solutions from the pareto-optimal set: a comparative study. In: Proceedings of the 9th Annual Conference on Genetic and Evolutionary Computation (GECCO 2007), pp. 789–796. ACM, New York (2007)

Cities as Enterprises: A Comparison of Smart City Frameworks Based on Enterprise Architecture Requirements

Viviana Bastidas[✉], Marija Bezbradica, and Markus Helfert

School of Computing, Dublin City University, Dublin, Ireland
{viviana.bastidasmelo, marija.bezbradica,
markus.helfert}@lero.ie

Abstract. There is a significant challenge in smart cities implementations. One challenge is to align smart city strategies with the impact on quality of life. Stakeholders' concerns are multiple and diverse, and there is a high interdependency and heterogeneity of technologies and solutions. To tackle this challenge, researchers have suggested to view cities as enterprises and apply an Enterprise Architecture (EA) approach. This approach specifies core requirements on business, information, and technology domains, which are essential to model architecture components and to establish relations between these domains. Existing smart cities frameworks describe different components and domains. However, the main domain requirements and the relations between them are still missing. This paper identifies essential requirements of enterprise architecture in smart cities. These requirements will be used to review and compare current smart city frameworks.

Keywords: Enterprise architecture · Requirement · Smart city framework

1 Introduction

A smart city is an ultra-modern urban area that addresses the needs of businesses, institutions, and citizens [1]. Smart city implementation assumes applying a city strategy into an effective improvement of quality of life for the citizens. This is a complex task because it involves different sectors, multiple stakeholders, high interdependency, cross-sectoral cooperation, inter-departmental coordination, and novel, dynamic, and interactive services [2]. Smart city should be composed by a well-constructed business plan [3], and a design of information and technology infrastructure to provide a platform for services integration. Industry and academy have proposed several frameworks to describe smart cities architectures in response to this complexity.

These smart city frameworks specify different layers of the architecture but it is often difficult to see a clear connection between these layers and the alignment with the smart city strategy. Some of these frameworks focus on city goals, objectives and indicators whereas others emphasize on solution architectures and technical details.

E. Alba et al. (Eds.): Smart-CT 2017, LNCS 10268, pp. 20–28, 2017.
DOI: 10.1007/978-3-319-59513-9_3

Enterprise architecture (EA) is the process of translating business vision and strategy into an effective enterprise change by creating, communicating and improving the key requirements, principles, and models that describe the enterprise's future state and enable its evolution [4]. The benefits of applying EA are visible in increased stability of an enterprise in an environment of constant change, better strategic agility, and improved alignment with business strategy. If a smart city is modelled as an urban enterprise [5], enterprise architecture approach can support its development and transformation based on the city strategic plan.

The Open Group Framework Architecture (TOGAF) specifies a detailed method for developing EA. This framework is based on interrelated areas of specialization called architecture domains: business architecture, information system architecture and technology architecture [6]. TOGAF specifies that the output of the previous domain is the input for the next one. It highlights that the relevance of each solution should be a result of the design and development of these architecture domains [6]. These domains have core requirements which are defined as entities in the TOGAF content metamodel. The content metamodel defines these architectural concepts to support consistency, completeness, traceability and relationship of components and layers in the enterprise architecture [7].

Implementing a smart city initiative does not only mean to reach technological success. It requires to articulate smart projects to specific initiative, such as: to deliver high quality e-services and to achieve outcomes desirable for the citizens [8]. This paper proposes to view a city as an urban enterprise to achieve alignment between strategy and smart city services solutions. This helps to support its development and transformation. In this regard, we aim to identify essential requirements of enterprise architecture in smart cities. These requirements will be used to review and compare current smart city frameworks.

This paper is structured as follows: Sect. 2 presents our research approach followed by the review of the content metamodel. Next, we inspect smart city frameworks to select the ones which proposed a more completed definition. Thereafter, selected types of architectures will be compared regarding core requirements. Finally, we discuss on the completeness of smart city frameworks in relation to EA core requirements.

2 Research Approach

In this paper, core requirements of TOGAF, are used to examine selected smart city frameworks. The content metamodel of TOGAF is reviewed to extract core architectural requirements for the business, information and technology domains. Next, smart city frameworks are selected. This selection is based on the major detail of description, number and type of layers. Then, the core requirements for each EA domain are mapped to the smart city frameworks. Finally, a discussion of the findings is presented in Sect. 6.

3 Enterprise Architecture Core Requirements

Enterprise architecture serves as a valuable instrument to guide the enterprise through the transformation from a current to a future state [9]. The scope of the enterprise architecture includes the people, processes, information and technology of the enterprise, and their relationships to one another [4]. TOGAF document specifies a detailed method and a set of supporting tools for developing Enterprise Architecture. Authors have chosen TOGAF as an example of the standard that combines business and technical threads [7].

TOGAF defines three architecture domains: business architecture, information systems architecture and technology architecture [6]. Business architecture describes the strategy plan of the services offered and structure, functions, resources and process involved. Information system architecture supports the information of business architecture and the concerns of the stakeholders. It includes the development of data and application architectures. Data architecture describes how data is collected, stored, arranged, integrated, used and governed in different data systems. Application architecture describes the set of applications, the interaction between them and the relation with the core business process. Technology architecture describes the hardware, software and network infrastructure required to support the implementation of data and applications.

These architecture domains have core requirements which are defined as entities in the TOGAF content metamodel. The TOGAF content metamodel introduces the relevant concepts of the architecture represented in entities and key relationships that support architectural traceability [7]. These entities incorporate most of the enterprise architecture components and the content metamodel shows the relations between them. Enterprise Architecture claims to align and integrate strategy, people, business and technology and for that reason, these relations are fundamental in an EA framework.

Table 1. Enterprise architecture core requirements adopted from the core content metamodel [6].

Enterprise architecture domain		Core Requirements (Entities TOGAF content metamodel)
Business architecture	Definition of business including strategy, structure and process	Goal
		Business process
		Organization unit
		Business services
Information system architecture	Data architecture, representation of all data systems	Data entity
	Application architecture, set of applications and their interfaces	Application portfolio
		Interface catalog
Technology architecture	The software and hardware infrastructure required to support the deployment of system components	Technology standards

This paper focuses on these entities viewed as core requirements in the architecture domains. This is defined in Table 1. The core requirements in the right column are described as follows:

- Smart city initiatives are based on a set of goals. For this reason, goals are selected as a core requirement.
- Business processes are the operational activities that provide, produce and deliver its business services. It is important to analyse the existence of the business process within the smart city context.
- Application portfolio and interface catalogue represent an encapsulation of application functionality aligned to implementation structure and the interfaces between smart city services respectively.
- Data entities define which data are being used by the business functions, processes, and smart services in the city. They can further model citizens, employees, innovators and visitors.
- The Technology standards document the agreed standards for technology across the smart cities, such as ubiquitous computing, big data, cloud computing, service-oriented architecture, IoT and the smart city technologies related.

4 Selecting Smart City Frameworks

A smart city framework enables cities to constitute a standard and a guide for an implementation and management of smart city services [10]. This paper includes research that describe frameworks for smart cities in a conceptual and architectural fashion. The criteria for selection encompasses the presence of at least three specific layers or domains, the detail of layers' description, and the existence of some business components in the frameworks. The review is based on the concept-centric approach for literature reviews [11]. In the next section, five frameworks for smart cities are reviewed.

4.1 Review of Smart City Frameworks

This section provides a high-level comparison and analysis of five smart city frameworks. Table 2 compiles and synthesizes how each research work meets core requirements from an EA point of view.

Cisco provides a conceptual Smart City Framework [12] that enables stakeholders to thrust and test smart city initiatives. The framework has four layers which consider the city´s objectives in social, environmental and economic terms; city indicators to measure and benchmark cities using predefined methodologies; city components related to city´s physical locations; and city content to encompass how smart city solutions are implemented. The Cisco smart city framework emphasizes the relationship among five key stakeholders: policy, regulators, developers, owners and operators. This framework presents a high-level overview of city objectives and their connection with city indicators. The city content layer deals with the development and

Table 2. Smart City Frameworks and EA Core Requirements

Core Requirements	Business architecture				Data architecture	Application architecture		Technology architecture
References	Goal	Business process	Organization unit	Business services	Data entity	Application portfolio	Interface catalog	Technology standards
Cisco Smart city framework: a systematic process for enabling smart + connected communities [12]	X		X					
Framework for smart city applications based on participatory sensing [13]			X		X	X		X
A conceptual enterprise architecture framework for smart cities: A survey based approach [14]						X	X	
An Information Framework for Creating a Smart City Through Internet of Things [15]					X	X	X	X
A Community Architecture Framework for Smart Cities. Citizen's Right to Digit City [16]	X		X			X		X

implementation of smart city solutions. This layer is connected to city components and objectives. However, this framework does not address layers and connections between data, application, and technology.

The Framework for Smart City Applications Based on Participatory Sensing [13] is constructed upon XMPP (Extensible Messaging and Presence Protocol) for mobile participatory sensing based on smart city applications. This framework proposes layers for mobility prediction: the streaming framework, persistence components, and custom analytics. This framework enables service innovation and the emerging popularity of crowd-sensing for data collection. Three roles are defined such as producers, service providers, and consumers which interact among them using an event based publish-subscribe mechanism. This framework also provides a value chain which presents details of the data collection and the interaction between the roles. However, this framework does not address completely the business layer with the specification of smart city goals, processes and business services.

A conceptual Enterprise Architecture Framework for Smart Cities [14] is proposed based on relevant quality properties of smart cities. This framework examines the current state in business aspects in relation to the IT support for smart city projects. It provides a justification through quality attributes using a layered approach rather than concentrating on specific types of services that may differ from city to city. This framework includes a business logical layer with diverse public or private services. The application layer uses the messaging pattern to gather data from different applications and interfaces. The proposed framework makes suggestions for the business aspects, however, it focuses only on the application layer of the smart city.

A framework for the realization of smart cities through the Internet of Things (IoT) [15] is presented as an urban information system, from the sensor level and networking support structure. This framework proposes an IoT infrastructure from three different domains: network-centric IoT for communication, cloud-centric IoT for management, and data-centric IoT for computation. This information framework involves networking modules and their connection with the communication stack (i.e. application, transport, network, MAC and physical layer). The data is interpreted, managed, processed and collected by the application layer. The integration is supported by data management and cloud-based integration. Even though this research presents a framework to model cost based on supply and demand, there are no additional details about components of the business layer such as goals, processes, organization units or business services.

A community Architecture Framework for Smart Cities [16] is presented to tackle the complexity that represents the management of multiple stakeholders, their inter-relationships and the conflict of interest resolution. Goals of each stakeholder, planner, public and developer are considered. This framework is used to develop a tool to support developers, planners, and communities to participate in the planning of their cities by proposing innovative ideas for their areas of interest. The research work is based on the Zachman framework [17]. The community architecture framework consists of data, function, network, organization, schedule and strategy components. The connection between these components is represented through different artifacts and models. However, the details of these artifacts or models are not present.

Table 2 indicates the main findings. Only few frameworks concentrate on the business layer (goals, objectives and city indicators). The majority are focused on data, application and technology layers (i.e. solution architecture and technical details). The next section presents a discussion of the findings which are related to the completeness of the smart city frameworks and the existence of a clear connection between different layers.

5 Discussion

The review allows to identify that various conceptual frameworks deal with smart city goals. However, they do not provide a deeper insight on data, application, and technology aspects. Some architectural frameworks comprise of stakeholders (i.e. organizational unit), data, application, and technical requirements. But, the relationship between different domains of the evaluated frameworks is still absent.

The evaluated frameworks are focused on smart city services and they do not consider the existence of business services. The last evaluated framework mentions community processes but there is not a major detail about these architectural components. The rest of the evaluated frameworks, do not explicitly take the business process into account. It is important to analyze in future works the relevance of business process within smart cities because smart cities focus on services.

Smart cities require linking smart projects with specific initiatives, such as: to deliver high-quality e-services, to achieve outcomes seen as desirable by the citizens and to increase trust in public institutions [8]. The lack of business plans in many cases, as well as the existence of a multitude of potential business goals, imply that these projects can be considered at the time as an umbrella under which many different applications coexist and can grow in many different directions in the future [14]. However, some evaluated frameworks do not involve a strategic plan based on goals.

Table 3. Research proposal and impact on quality of life

Research proposal	Impact
Achieving alignment between smart city strategy and services	Delivering high-quality services connected to real needs of citizens
	Offering better and more convenient services for citizens
	Understanding service requirements to deliver desirable outcomes
Connecting smart city solutions and initiatives (i.e. education, health, transportation, environment initiatives, etc.)	Supporting development and transformation of sustainable cities
	Delivering benefits to citizens to improve their quality of life

This paper proposes to view a city as an urban enterprise to achieve alignment between strategy and smart city services. Smart city implementations require applying the city strategy into an effective improvement of quality of life for citizens. The impact of this approach and the different alignment levels are indicated in Table 3.

6   Conclusions and Further Research

Many cities are experiencing exponential growth as people move from rural areas looking for better jobs and services. Consequently, cities' services and infrastructures are being stretched to their limits in terms of scalability, environment, and security as they adapt to support this population growth [1]. There is a significant challenge in the implementation of Smart City Transformation. Understanding cities as enterprises, we can apply Enterprise Architecture [18]. This approach can support the development and transformation of Smart Cities based on the strategic plan of the city and the needs of the citizens. Enterprise Architecture can help to establish the current and the desired state of the city.

Existing smart cities frameworks describe different components and domains. This paper identifies essential requirements of enterprise architecture in smart cities. These requirements are used to review and compare current smart city frameworks. The obtained results indicate that few smart city frameworks are concentrated in the business layer and the majority are focused on data, application, and technology architecture; and there is no a clear connection between these domains. These frameworks have their own advantages in terms of addressing smart cities challenges, however, this comparison does not include technical issues or technical details to indicate possible success or failure of these frameworks.

As further steps for this research, it is important to define the connection between layers in an Urban IT Reference Architecture and to define use cases in real scenarios to validate this approach. Moreover, smart city frameworks can be explored based on different EA domains and the TOGAF metamodel.

Acknowledgments. This work was supported by the Science Foundation Ireland grant "13/RC/2094" and co-funded under the European Regional Development Fund through the Southern & Eastern Regional Operational Programme to Lero - the Irish Software Research Centre (www.lero.ie).

References

1. Khatoun, R., Zeadally, S.: Smart cities: concepts, architectures, research opportunities. Commun. ACM **59**(8), 46–57 (2016)
2. Javidroozi, V., Shah, H., Amini, A., Cole, A.: Smart city as an integrated enterprise: a business process centric framework addressing challenges in systems integration. In: Third International Conference on Smart Systems, Devices and Technologies (SMART 2014), pp. 55–59 (2014)

3. Pourzolfaghar, Z., Bezbradica, M., Helfert, M.: Types of IT architectures in smart cities – a review from a business model and enterprise architecture perspective. In: AIS Pre-ICIS Workshop on "IoT & Smart City Challenges and Applications" (ISCA 2016), 9 December 2016, Dublin (2016)
4. Lapkin, A., Allega, P., Burke, B., Burton, B., Bittler, R.S., Handler, R.A., James, G.A., Robertson, B., Newman, D., Weiss, D., Buchanan, R., Gall, N.: Gartner clarifies the definition of the term enterprise architecture. In: Gartner Research (2008)
5. Mamkaitis, A., Bezbradica, M., Helfert, M.: Urban enterprise: a review of smart city frameworks from an enterprise architecture perspective. In: Proceedings of the IEEE 2nd International Smart Cities Conference. Improving the citizens quality of life (ISC2 2016), pp. 1–9 (2016)
6. The Open Group: Open Group Standard TOGAF Version 9.1. (2011)
7. Czarnecki, A., Orłowski, C.: IT business standards as an ontology domain. In: Jędrzejowicz, P., Nguyen, N.T., Hoang, K. (eds.) ICCCI 2011. LNCS, vol. 6922, pp. 582–591. Springer, Heidelberg (2011). doi:10.1007/978-3-642-23935-9_57
8. Dameri, R.P.: Searching for smart city definition: a comprehensive proposal. Int. J. Comput. Technol. **11**, 2544–2551 (2013)
9. Meyer, M., Helfert, M., O'Brien, C.: An analysis of enterprise architecture maturity frameworks. In: Grabis, J., Kirikova, M. (eds) Perspectives in Business Informatics Research, BIR 2011. LNBIP, vol. 90, pp. 167–177. Springer, Heidelberg (2011)
10. British Standards Institution (BSI): Smart city framework – guide to establishing strategies for smart cities and communities. BSI Stand. Publ. PAS 181 (2014)
11. Webster, J., Watson, R.T.: Analyzing the past to prepare for the future: writing a literature review. MIS Q. **26**, 13–23 (2002)
12. Falconer, G., Mitchell, S.: Smart city framework: a systematic process for enabling smart + connected communities, p. 11. Cisco Internet Business Solutions Group (IBSG), San Jose (2012)
13. Szabo, R., Farkas, K., Ispany, M., Benczur, A.A., Batfai, N., Jeszenszky, P., Laki, S., Vagner, A., Kollar, L., Sidlo, C., Besenczi, R., Smajda, M., Kover, G., Szincsak, T., Kadek, T., Kosa, M., Adamko, A., Lendak, I., Wiandt, B., Tomas, T., Nagy, A.Z., Feher, G.: Framework for smart city applications based on participatory sensing. In: Proceedings of the 4th IEEE International Conference on Cognitive Infocommunications (CogInfoCom 2013), pp. 295–300 (2013)
14. Kakarontzas, G., Anthopoulos, L., Chatzakou, D., Vakali, A.: A conceptual enterprise architecture framework for smart cities: a survey based approach. In: 11th International Conference on e-Business (ICE-B 2014) - Part 11th International Joint Conference on e-Business Telecommunications (ICETE 2014), pp. 47–54 (2014)
15. Jin, J., Gubbi, J., Marusic, S., Palaniswami, M.: An information framework for creating a smart city through Internet of Things. IEEE Internet Things J. **1**, 112–121 (2014)
16. Ilhan, A., Möhlmann, R., Stock, W.G.: A community architecture framework for smart cities. In: Foth, M., Brynskov, M., Ojala, T. (eds.) Citizen's Right to the Digital City, pp. 231–252. Springer, Singapore (2015)
17. Zachman, J.A.: A framework for information systems architecture. IBM Syst. J. **38**, 454–470 (1999)
18. Mamkaitis, A., Bezbradica, M., Helfert, M.: Urban enterprise principles development approach: a case from a European City. In: AIS Pre-ICIS Workshop on "IoT & Smart City Challenges and Applications", Dublin, pp. 1–9 (2016)

Comparative Study of Artificial Neural Network Models for Forecasting the Indoor Temperature in Smart Buildings

Sadi Alawadi[✉], David Mera, Manuel Fernández-Delgado,
and José A. Taboada

Centro Singular de Investigación en Tecnoloxías da Información (CiTIUS),
Universidade de Santiago de Compostela, Rúa de Jenaro de la Fuente Domínguez,
15782 Santiago de Compostela, Spain
saadiabadi@gmail.com

Abstract. The implementation of efficient building energy management plans is key to the road-map of the European Union for reducing the effects of the climate change. Firstly, accurate models of the currently energy systems need to be developed. In particular, simulations of Heating, Ventilation and Air Conditioning (HVAC) systems are essential since they have a relevant impact in both energy consumption and building comfort. This paper presents a comparative of four different machine learning approaches, based on Artificial Neural Networks (ANNs), for modeling an HVAC system. The developed models have been tuned to forecast three consecutive hours of the indoor temperature of a public research building. Tests revealed that an on-line learning ANN, which is also fully trained weekly, is less affected by sensor noise and anomalies than the remaining approaches. Moreover, it can be also automatically adapted to deal with specific environmental conditions.

Keywords: Smart buildings · Time series prediction · Energy efficiency · Neural network

1 Introduction

The European Union (EU) is currently supporting several initiatives focused on reducing the effects of the climate change. These promoted plans have several objectives such as the reduction of the emissions of the member countries and the improvement of the energy efficiency. In particular, the EU, via the "Climate Action", has set out to decrease EU greenhouse gas emissions by at least 20%

This work has been developed under Erasmus Mundus Action 2, Strand 1, Lot 2, PEACE II, with project code 2013-2443/001-001. It has been partially supported by EC under the LIFE12- ENV-ES-001173 project application from Life+ Environment Policy and Governance Programme as well as by the Consellería de Cultura, Educación e Ordenación Universitaria (accreditation 2016-2019, ED431G/08) and the European Regional Development Fund (ERDF).

© Springer International Publishing AG 2017
E. Alba et al. (Eds.): Smart-CT 2017, LNCS 10268, pp. 29–38, 2017.
DOI: 10.1007/978-3-319-59513-9_4

and 40% below 1990 levels by 2020 and 2030, respectively. Moreover, it has also posed to improve the energy efficiency for reducing the amount of primary energy used by 20% and 27% in 2020 and 2030, correspondingly [1].

The efficient management of buildings has been pointed out as key to get the proposed objectives due to the fact that the EU estimates that buildings represent the 40% of energy consumption and the 36% of total CO_2 emissions within the Union [2].

The Universidade de Santiago de Compostela (USC), in the framework of the "Climate Action" initiative, has developed the European Opere project [3] with the objective of improving the USC energy management systems. This project allowed us to deploy a sensor network in 45 university buildings that generates more than 10,000 signals. In particular, our research interest is focused on the Centro Singular de Investigación en Tecnoloxías da Información (CiTIUS), which is a research building with a medium-size sensor network that produces 667 signals each 10 s. This raw dataset symbolizes a rich source of information that can be used to improve the energy efficiency, to detect system faults, to optimize the resources, etc.

The development of current building energy models is essential to establish improved energy plans since they would allow us to simulate the effect of new configurations. In particular, the simulation of Heating, Ventilation and Air Conditioning (HVAC) systems is indispensable to deal with these initiatives [4] due to the fact that they have a relevant influence in both the energy consumption and the user comfort. Concretely, HVAC systems have the highest energy demand in buildings [5] and they generate around 33% of global greenhouse gas emissions [4]. From the viewpoint of the user comfort, it must be noted that the impact of the different HVAC configurations is not immediately noted. Thus, an appropriate HVAC model should take also into account the future comfort behavior of the related facilities. Furthermore, the model also should consider external features such as weather conditions since they have influence in both the indoor temperature and the feeling of comfort.

Previous studies have shown that Machine Learning (ML) algorithms can be used to model building energy systems in general and HVAC systems in particular [6]. Artificial Neural Networks (ANNs) were used to achieve cooling load forecasting in HVAC systems in [7]. Doukas et al. developed a decision support system focused on improving an energy management system [8]. The influence of the building occupancy in a HVAC model was studied in [9] through Markov Chains. Recently, Rodríguez-Mier et al. shown that a genetic fuzzy algorithm can be successfully used to develop a rule-based model to predict the indoor building temperature [10].

Typically, ML algorithms are trained and tested with historical data. However, the performance of HVAC systems is closely related to external conditions such as weather. It would be desirable to have a large training set from different years linked to different weather conditions in order to develop an accurate model. Due to the fact that this is hard to obtain, adaptive ML methods seem to be appropriate approaches for dealing with this problem. Nevertheless, they can fall under the influence of stationary weather features, outliers, sensor noise, etc.

In this paper, we have developed four ML models based on ANNs to model the HVAC system in one of the CiTIUS offices: (1) a Multilayer Perceptron (MLP) network, that was trained with historical data; (2) an Adaptive Multilayer Perceptron (AMLP), which was iteratively trained with incremental datasets every week; (3) an Online learning Multilayer Perceptron (OMLP), that was updated after processing each pattern; and (4) an Online learning Adaptive Multilayer Perceptron (OAMLP) model, which was updated after processing each pattern and also fully trained with incremental datasets every week. The common goal of these models is to forecast 3 h of the indoor temperature in a building office. The objective of this study is to compare the forecasting accuracy of these models in a real environment, so we have taking into account both different weather conditions than the used for training the models and different noise sensor levels.

The remainder of this paper is organized in the following manner. Section 2 describes the dataset used in the experiments. The second section is a brief introduction to the ANNs algorithms. Section 4 describes the developed ANN approaches as well as the experimental setup. Results and discussion is shown in Sect. 5. Finally, Sect. 6 draws the conclusions and outlines the future work.

2   Data Acquisition

A dataset composed of both sensor measurements linked to the CiTIUS HVAC system and weather data from the closest Meteogalicia weather station was used to develop the experiments described in this manuscript. Specifically, the dataset patterns were obtained every 10 min during two time periods: from 1st October 2015 to 31st March 2016 (26,321 patterns) and from 1st November 2016

Table 1. Pattern features, where (*) represents features from CiTIUS, and (+) symbolizes features from Meteogalicia.

Features	Abbr.	Type	Description
Underfloor heating status*	UHS	Binary	Status of the underfloor heating system in the office
Underfloor heating temperature*	UHT	Continuous	Temperature of the water linked to the underfloor heating system
Air condition status*	ACS	Binary	Status of the air conditioning system in the office
Air conditioning temperature*	ACT	Continuous	Temperature of the main air conditioning system
Air conditioning humidity*	ACH	Continuous	Percentage of the humidity linked to the main air conditioning system
Humidity+	OutH	Continuous	Degree of the outdoor relative humidity
Temperature+	OutT	Continuous	Outdoor temperature
Solar radiation+	SR	Continuous	Level of solar radiation
Indoor temperature*	T	Continuous	Indoor temperature related in the office
Previous indoor temperature*	T − 1	Continuous	Office indoor temperature at the instant T − 1

to 31st January 2017 (13,083 patterns). Both periods correspond to the HVAC winter working mode, which has the highest energy demand. It must be noted that the second period corresponds to an unusual dry winter season in Galicia. Thus, the weather conditions in both periods are different enough. The datasets used in this study are available on-line[1], so that the experiments can be repeated if necessary.

Each dataset pattern is made up of 10 features, which are described in Table 1.

3   Artificial Neural Networks

ANNs have become one of the most popular ML methods for learning from data. They try to simulate the biological neural networks for solving complex problems. Basically, an ANN learns how to map input data (patterns) onto appropriate outputs. The generalization capability is one of the most important ANN features, which means that they can analyze unseen patterns and obtain convenient outputs for them. ANNs are composed of simple units called neurons, which are interconnected to cooperate. Briefly, each neuron gets a set of inputs, combines the input values together and applies an activation function over them for obtaining the neuron output. Typically, a sigmoid or a hyperbolic tangent function is used as the activation function. Connections between neurons can be enforcing or inhibitory through linked weights to get the desired output. A set of neurons that obtain all the input connections from the same source set and connect all their outputs to the same target set is called layer. There are different types of ANNs. In this work we have used MLPs, which are feed-forward networks composed of several neuron layers, where each layer is fully connected to the next one in a direct graph. Typically, a MLP is made up by an input layer, that obtains external data, several processing hidden layers and an output layer, which produces network outputs. A MLP scheme is shown in Fig. 1. A detailed description about ANN and their different types can be found in [11].

4   Materials and Methods

We have developed four ML algorithms based on ANNs to model the CiTIUS HVAC system. Concretely, these models were tuned to forecast three hours of the indoor temperature in a CiTIUS office. We used different ways to train them in order to compare their accuracies in a real testing environment that is affected by different weather conditions and noise levels.

1. MLP: this is the simplest approach since it was trained with a fixed historical dataset and it was independently tested with new data. This model cannot be updated in runtime.

---

[1] https://gitlab.citius.usc.es/cograde/HVAC-model.

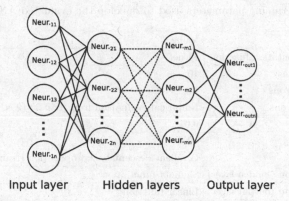

Input layer   Hidden layers   Output layer

Fig. 1. Simplified scheme of a Multilayer Perceptron network

2. AMLP: this approach was initially trained with historical data. However, the AMLP was iteratively trained and adapted as it was processing new data. Concretely, this approach was trained every week with an incremental dataset that includes both weekly data and old data. Thus, the AMLP can be adapted to new scenarios. It must be noted that the network architecture can be changed in each training iteration.
3. OMLP: this approach was initially trained with historical data. After that, it was updated with every new pattern for learning faster. It must be noted that there was a delay between the temperature prediction and the use of the related pattern to update the network due to the fact that it was necessary to wait for the real indoor temperature linked to that pattern.
4. OAMLP: this approach was a combination between OMLP and AMLP. OAMLP was updated with each new pattern and it was also trained with an incremental dataset once a week as was previously described.

4.1 Experimental Setup

We developed the evaluated approaches using the dataset described in Sect. 2. In particular, we used the first period (from 1st October 2015 to 31st March 2016) to both train and validate the models and the second period (from 1st November 2016 to 31st January 2017) to test them. This allowed us to evaluate the models with different weather conditions. Moreover, we have also artificially generated other test set alternatives adding noise to the original one in order to evaluate different noise scenarios. Train and validation partitions were randomly generated in such a way that 85% of the patterns were used for training the models and the remainder for validating them. The evaluated ANN models were implemented using Matlab. Taking in mind a fair comparison, all of them were tuned with the same parameters during the training phase. A detailed list of them is shown in Table 2 and it is also available on-line (see Footnote 1).

All the developed models were composed of three layers: input layer, hidden layer and output layer. The number of neurons linked to the hidden layer

Table 2. Training parameters used to develop the evaluated ANN models

Parameter	Value
Network implementation function	Fitnet
Input layer neurons	10
Output layer neurons	1
Hidden layer neurons	Empirical evaluation from 1 to 20 (2 × input_neurons)
Learning rate	0.01
Epochs	1000
Training function	Gradient descent backpropagation (traingd)
Activation function (hidden layer)	Sigmoid function
Activation function (output layer)	Linear function
Divide function	Divideind (index are available on-line (see Footnote 1)
Divide mode	Sample

was fixed for each approach through empirical tests based on the 'ad hoc' rule that this number should not be higher than the double of neurons of the input layer [12]. Each architecture was validated using the validation set. Finally, for each approach, the architecture with the best performance was selected for the final model.

The AMLP, OMLP and OAMLP approaches were developed on the basis of the trained MLP model. Due to the fact that they could be adapted, these approaches had a specific training process. The AMLP approach was iteratively trained every week using both weekly data (1,008 patterns) and historical data. Weekly patterns were distributed in each iteration between the train and validation sets (80% and 20%, respectively). These new enriched datasets were used to train and adapt the MLP model. The OMLP approach was updated at runtime using the input patterns once the real indoor temperature was available. The architecture of this network remained unchanged during this process. Finally, the OAMLP approach was a hybrid model between the AMLP and the OMLP since it was updated at runtime and also trained and adapted once a week.

The noise sensitivity of the developed ANN models was examined adding different noise levels (5%, 10%, 15%, 20% and 25%) to the test set (see Footnote 1), in order to simulate noise and temporary failures related to the sensors. Specifically, this noise was added using the MATLAB Gaussian noise function (awgn).

We repeated the experiments 10 times for each model using the same dataset but using different seeds (from 1 to 10) for generating different partitions and network initializations. After that, we averaged the results for each model.

5   Results and Discussion

As a case-study, we have focused on the 3-hours indoor temperature forecast of a CiTIUS office. Figure 2 plots the results of the four tested approaches. In

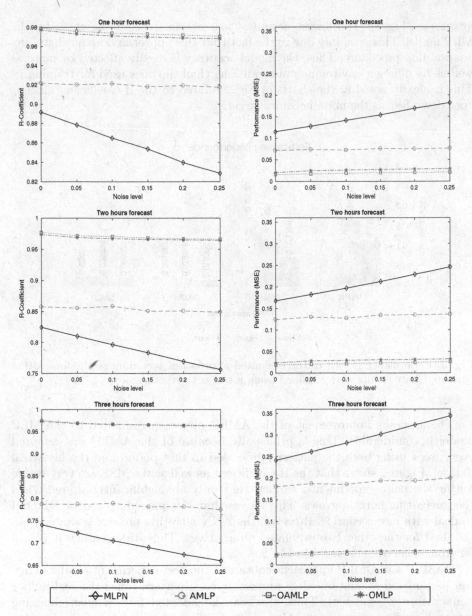

Fig. 2. R-Coefficient and MSE metrics of the evaluated approaches according to the forecasting horizon. Left panel: R-coefficient measurements. Right panel: MSE values.

particular, we have evaluated them by computing both the R-Coefficient and the performance. Specifically, the latter was calculated via the Mean Square Error (MSE). Results show that the approaches obtained a reasonable accuracy in all the evaluated scenarios. Obviously, it was worse as the forecasting horizon grows (see Fig. 3). The lowest accuracy was obtained in all the cases by the MLP model. This is mainly due to the fact that this approach does not have any adaptability procedure. Thus, the model accuracy is hardly affected by noise as well as by different environmental conditions than the ones used for training it. This is clearly noted in the charts of Fig. 2 related to the R-Coefficient since it goes down fast as the noise becomes larger.

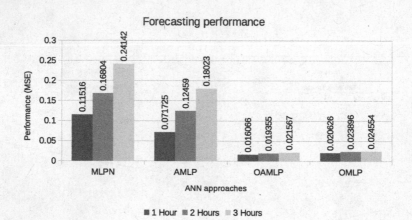

Fig. 3. Performance (MSE) of the evaluated approaches according to the forecasting horizon. Chart results were obtained using a test set without noise

The accuracy improvement of the AMLP approach compared to the MLP model is considerable. This is principally because of the AMLP was adapted every week using both the patterns processed in that period and the historical dataset. Figure 2 shows that the R-Coefficient as well as the MSE are very stable during the noise experiments. In fact, they only are significantly altered when the forecasting horizon grows. This means that this procedure can be adapted to deal with new scenarios. However, the ANN adapting process is slower than the used for the other two evaluated alternatives. Thus, its accuracy it is still far from them as is depicted in Fig. 2.

OAMLP and OMLP approaches obtained the best experiment results. They can be updated at runtime after processing each pattern (once the real indoor temperature is available). Thus, they can be adapted fast to both changing external conditions and noise. Consequently, they obtained better forecasting accuracy as well as the highest metrics in Fig. 2. They had similar R-coefficient metrics in all the scenarios. However, the performance charts show that the OAMLP approach obtained slightly better results. We consider that this is due to the fact that the OMLP is updated with every pattern, so it can fall under the influence of outliers and anomalies. However, the OAMLP is trained every week. Thus, model deviations caused by anomalies and outliers can be fixed.

It must be highlighted that both models OAMLP and AMLP can automatically change their architecture in each training iteration. Table 3 shows the

Table 3. Change neurons numbers in the hidden layer in both lambda and lambda-adapt models during the periodical training for one index

Noise %	Iteration											
	1	2	3	4	5	6	7	8	9	10	11	12
0	6	6	6	6	6	6	6	6	6	16	6	6
5	6	7	6	6	16	6	6	6	6	6	6	6
10	6	6	6	6	6	16	6	16	6	6	6	6
15	7	7	6	6	6	6	6	16	6	16	6	6
20	6	6	6	6	6	6	6	6	6	6	6	6
25	6	6	6	6	6	6	6	6	16	16	6	6

different architectures used during the experiments. Despite 6 neurons is the most used architecture, some iterations required a different number of neurons. We consider this a relevant characteristic since it allows models to be automatically improved as more data are obtained.

6 Conclusion and Future Work

Four different ML approaches based on ANNs were developed to model the CiTIUS HVAC system and to forecast the indoor temperature of a CiTIUS office. These approaches were: MLP, OMLP, AMLP and OAMLP. The accuracy of these approaches was studied and compared in this work taking into account different weather conditions and sensor noise levels.

According to the obtained results, an adaptive approach is essential to model an HVAC system and to forecast the indoor temperature of the linked facility. Simplest models like the regular MLP cannot deal with new environmental conditions and different noise levels. Nevertheless, adaptive approaches such as AMLP, OMLP and OAMLP can be automatically updated to different scenarios. However, we must take in mind that the use of an on-line approach like OMLP can fall under the influence of outliers. Test shown that a hybrid approach between an on-line learning procedure and an incremental weekly training process obtain the best metrics and it is more stable. On the one hand, it can be fast updated to new scenarios even when they are temporary. On the other hand, the deviations caused by short-lived events, outliers and model anomalies can be fixed when the network is trained once a week. Moreover, this approach can also change its architecture during the iterative training process. This allows to improve the model as more information is obtained.

We are currently working in the deployment of this model in all the CiTIUS facilities in order to get a full system where we would be able to test different HVAC configurations and to evaluate their effects. After that, we will develop an algorithm to automatically optimize the HVAC configuration.

References

1. Europese Commissie: A roadmap for moving to a competitive low carbon economy in 2050. Europese Commissie, Brussel (2011)
2. Zhao, H.-X., Magoulès, F.: A review on the prediction of building energy consumption. Renew. Sustain. Energy Rev. **16**(6), 3586–3592 (2012)
3. Life-OPERE. http://www.life-opere.org/en. Accessed Jan 2017
4. Kharseh, M., Altorkmany, L., Al-Khawaj, M., Hassani, F.: Warming impact on energy use of HVAC system in buildings of different thermal qualities and in different climates. Energy Convers. Manag. **81**, 106–111 (2014)
5. Fong, K.F., Hanby, V.I., Chow, T.-T.: HVAC system optimization for energy management by evolutionary programming. Energy Build. **38**(3), 220–231 (2006)
6. Dounis, A.I., Caraiscos, C.: Advanced control systems engineering for energy and comfort management in a building environmenta review. Renew. Sustain. Energy Rev. **13**(6), 1246–1261 (2009)
7. Beghi, A., Cecchinato, L., Rampazzo, M., Simmini, F.: Load forecasting for the efficient energy management of HVAC systems. In: 2010 IEEE International Conference on Sustainable Energy Technologies (ICSET), pp. 1–6. IEEE (2010)
8. Doukas, H., Patlitzianas, K.D., Iatropoulos, K., Psarras, J.: Intelligent building energy management system using rule sets. Build. Environ. **42**(10), 3562–3569 (2007)
9. Erickson, V.L., Carreira-Perpiñán, M., Cerpa, A.E.: OBSERVE: occupancy-based system for efficient reduction of HVAC energy. In: 2011 10th International Conference on Information Processing in Sensor Networks (IPSN), pp. 258–269. IEEE (2011)
10. Rodrıguez-Mier, P., Fresquet, M., Mucientes, M., Bugarın, A.: Prediction of indoor temperatures for energy optimization in buildings (2016)
11. López, R.F., Fernandez, J.M.F.: Las redes neuronales artificiales. Netbiblo (2008)
12. Swingler, K.: Applying Neural Networks: A Practical Guide. Morgan Kaufmann, San Francisco (1996)

Considering Congestion Costs and Driver Behaviour into Route Optimisation Algorithms in Smart Cities

Pablo Alvarez[✉], Iosu Lerga, Adrian Serrano, and Javier Faulin

Department of Statistics and OR, Institute of Smart Cities,
Public University of Navarra, Pamplona, Spain
{pablo.alvarez,iosu.lerga,adrian.serrano,
javier.faulin}@unavarra.es

Abstract. Congestion costs have been excluded from the study of traditional vehicle routing problems until very recently. However, with our urban areas experiencing higher levels of traffic congestion, with the increase in on-demand deliveries, and with the growth of intelligent transport systems and smart cities, researchers are raising awareness on the impact that traffic congestion and driver behaviour has for urban logistics. This paper studies the evolution of the vehicle routing problem, focusing on how traffic congestion costs and driver behaviour effects have been considered so far, and analysing how the research community has to deal with this challenge.

Keywords: Vehicle routing problem · Congestion · Driver behaviour · Smart cities · Big Data

1 Introduction

The Vehicle Routing Problem (VRP) is one of the recurring topics existing in the literature on transport and logistics activities [1]. In 1954, Dantzig, Fulkerson, and Johnson wrote a seminal paper [2] to solve a large-scale traveling salesman problem (TSP). However, it was in 1959 when Dantzig and Ramser [3] developed the first algorithmic approach applied to optimise delivery routes between petrol stations. This was the first record related to a VRP, as we know it today, although it was later improved (in 1964) by Clarke and Wright [4] to solve a problem in which a fleet of trucks of different capacities had to be used for delivery from a central depot to several delivery points. A few years later, different versions of VRP emerged and were applied to different fields such as fleet routing [5], bus routing [6], or waste collection [7]. However, it was not until 1972 when the words *"vehicle routing"* appeared together in the title of a research work by Golden, Magnanti and Nguyen [8], in which they used heuristic programming to develop a multi-depot routing algorithm. In that period, researchers tried to solve the problems in a more realistic way, using more constraints such as vehicle capacity, time windows, or different fleet configurations.

The possibilities to study more realistic variations of VRP were limited due to the computational complexity required, and it was not until the 1990s when research on

© Springer International Publishing AG 2017
E. Alba et al. (Eds.): Smart-CT 2017, LNCS 10268, pp. 39–50, 2017.
DOI: 10.1007/978-3-319-59513-9_5

VRP accelerated mainly due to microcomputer capability and availability, which allowed the introduction of meta-heuristics into the study of VRP applications.

The VRP field has greatly evolved since those first studies written decades ago, and hundreds of new papers on more complex variants of the VRP are being released every year. These new approaches take into account new constraints such as the stochastic effects of real-life problems by using simheuristics [9], the range limitations when delivering with electric vehicles and even drones, or the changing conditions existing in the market by considering real-time information to enhance dynamic vehicle routing problems.

Nevertheless, the scale at which VRP are being used has not change much in more than 55 years, as the majority of applications are still focused at a strategic level (i.e. routes between cities) where most of the approaches aim to minimise a distance-based cost function without considering other aspects such as congestion effects, transport infrastructure, or driver behaviour.

By 2050, 70% of the world's population is predicted to live in cities [10], putting increased strain on urban infrastructure and transport systems. Adding to that the fact that we live in the one-click Era, where online shopping and on-demand deliveries are changing the way we understand logistics activities within urban areas, it is easy to understand that the way we optimise urban logistics routes must change too. Making use of Big Data technologies, traffic simulation software, and considering dynamic local conditions such as congestion effects or driver behaviours is necessary to get realistic solutions to our optimisation problems.

Therefore, this paper aims to study how current VRP approaches are suitable for their use in urban logistics, specifically focusing on how congestion effects and driver behaviours are being considered. To do this, the latest published papers on city logistics, which include congestion and driver behaviour effects, have been analysed. The advantages and the disadvantages of the different approaches have been also described to understand how convenient they are for their use in real urban scenarios. Finally, the authors draw some conclusions on how future research on this topic should addressed.

2   Methodology

The following methodology has been applied in order to analyse how appropriate current approaches are when considering congestion effects and driver behaviours into VRP.

2.1   Number of Indexed Publications per Year

Using Scopus, a search has been performed to check the number of indexed publications per year related to (1) "Vehicle Routing Problem", (2) "Vehicle Routing Problem" AND "logistics", (3) "Vehicle Routing Problem" AND "logistics" AND "congestion", (4) "Vehicle Routing Problem" AND "logistics" AND "driver behaviour", (5) "Urban logistics", and (6) "Logistics" AND "smart cities".

Through this, it is possible to get a broad picture of how VRP and logistics have evolved, and how relevant aspects such as congestion costs and driver behaviour have been for researchers.

2.2 Literature Overview

A comprehensive analysis of all the papers related to VRP (focused on logistics) in which congestion and driver behavioural aspects are considered has been performed. A table has been created in which the following attributes are scored.

- **Title**
- **Author**
- **Year**
- Relevance (**R**) of congestion effects or driver behaviour within the whole paper. From 0 (no relevant at all) to 5 (extremely relevant).
- Originality (**O**). How original is the approach used to consider congestion/driver behaviour effects within the VRP field? From 0 (no original at all) to 5 (extremely original).
- Practicality (**P**). How practical/realistic is the approach used to consider congestion or driver behaviour within the VRP field? From 0 (no practical at all) to 5 (the approach is completely realistic and practical).
- Age (**A**). How new is the paper? The more up-to-date a paper is, the more likely it is that it includes aspects related to cutting-edge technologies or approaches. The value is 5 (2017 or 2016), 4 (2015 or 2014), 3 (2013 or 2012), 2 (2011 or 2010), 1 (2009 or 2008) or 0 (before 2008). If R = 0, then A = 0.
- Continuity (**C**). Does the paper lay the foundation for future research? The value goes from 0 (the paper has no continuity) to 5 (it is very likely that more papers are written based on this one).
- **ROPAC value**. It is the sum of the paper's relevance (R), originality (O), practicality (P), age (A), and continuity (C). The maximum value is 25, and the minimum is 0. The higher the value is, the more important the paper is for the field.

2.3 Analysis of the Five Papers with the Highest ROPAC Value

The five papers with the highest ROPAC values will be described, focusing on how congestion costs or driver behaviour effects have been taken into account.

3 Analysis of Results

As commented in Sect. 2, the analysis has been split into three main parts: number of indexed publications per year (evolution), general literature overview, and analysis of the five papers with the highest ROPAC value.

3.1   Number of Indexed Publications per Year

Figure 1 shows the evolution of the research done in the field of vehicle routing problems in the last decades. Note that just Scopus-indexed papers are considered.

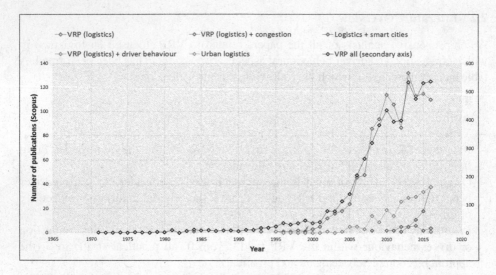

Fig. 1. Evolution of VRP, and urban logistics within smart cities.

The first indexed papers related to VRP were published in the 1970s. In fact, the first one was a piece of research [11] in which the authors developed an approach to optimise public service vehicle routes. During the next twenty years, the knowledge on this field was consolidated, and it was not until the 1990s when the number of publications started to increase exponentially. It was also then when specific papers focused on VRP for logistic activities started to be indexed (although Daganzo published the first one [12] in 1984).

Until 1995, research on VRP was mostly focused on a strategic (macroscopic) level, but the research community commenced to adopt a more micro-level perspective and the study on urban logistics started to be considered. Since then, research on this field has greatly increased, and it has been complemented by publications on logistics and smart cities that began to be indexed in 2012. Nowadays, the number of indexed publications related to VRP reaches more than 500 papers per year. For VRP focusing on logistics, the approximate figure drops to 110 per year, following the same tendency as general VRP publications. With urban logistics and smart cities gaining in prominence during the last years, the number of publications has increased to 40 indexed papers by year.

However, it is shown in the figure that VRP papers focused on logistics have ignored congestion and driver behaviour effects until recently, when urban logistics and smart cities started to be considered by different authors, and it was not until 2008 when the first indexed paper appeared [13] to study a VRP that includes queues and dynamic travel times. Note that there is just one indexed publication (a survey from 2016) in which the driver behaviour aspects are mentioned [14].

Therefore, it is clear that congestion and driver behaviour effects within VRP have been excluded from the literature until now, and even with the current increase in the number of indexed papers published related to urban logistics and smart cities, these aspects are being disregarded.

3.2   General Literature Overview

All the papers indexed in Scopus related to vehicle routing problem for logistics activities in which congestion and/or driver behavioural effects are mentioned have been analysed according to the methodology stated in Sect. 2. Table 1 shows the scores for each paper, which have been agreed by all the authors.

Table 1. Scores of each paper.

Paper	Year	R	O	P	A	C	ROPAC	Paper	Year	R	O	P	A	C	ROPAC
[15]	2016	5	3	4	5	4	21	[16]	2013	2	3	2	3	2	12
[17]	2017	4	4	4	5	4	21	[18]	2012	2	3	2	3	2	12
[19]	2014	4	4	3	4	4	19	[20]	2014	2	2	2	4	1	11
[21]	2012	4	4	4	3	4	19	[22]	2015	1	2	1	4	1	9
[23]	2010	5	4	4	2	4	19	[24]	2013	2	1	1	3	1	8
[25]	2012	3	4	4	3	4	18	[26]	2013	2	1	1	3	1	8
[27]	2014	3	3	3	4	3	16	[28]	2008	3	1	2	1	1	8
[29]	2012	4	3	3	3	3	16	[30]	2011	2	1	1	2	1	7
[13]	2008	4	3	4	1	4	16	[31]	2012	2	1	0	3	0	6
[32]	2017	3	2	2	5	3	15	[33]	2014	0	0	0	0	0	0
[34]	2015	3	3	3	4	2	15	[35]	2014	0	0	0	0	0	0
[36]	2016	2	3	2	5	2	14	[37]	2014	0	0	0	0	0	0
[14]	2016	5	1	1	5	2	14	[38]	2010	0	0	0	0	0	0
[39]	2013	2	3	3	3	3	14	[40]	2016	-	-	-	-	-	-
[41]	2011	3	2	2	2	4	13								

The maximum ROPAC score (21) is for the paper written by Kim et al. in 2016 [15] which is called *"Solving the dynamic vehicle routing problem under traffic congestion"*, and for the one [17] by Huang et al. *"Time-dependent vehicle routing problem with path flexibility"* from 2017. On the other hand, there are four articles scoring zero, being the oldest one the paper [38] written by Danielis in 2010 on *"Urban freight policies and distribution channels"*. One of the indexed papers [40] was not accessible to the authors, therefore it was not analysed. The average ROPAC values is 11.41, and the median is 13.

3.3   Analysis of the Five Papers with the Highest ROPAC Value

The five papers with the highest ROPAC value are described. Although more papers with high ROPAC values could be analysed to give a wider vision, it has been checked that no decisive aspects have been missed when excluding the rest of the papers from this analysis.

3.3.1   Algorithms to Quantify Impact of Congestion on Time-Dependent Real-World Urban Freight Distribution Networks [23]

Through a real case study in Portland, Oregon, the authors make use of two main data sources. On the one hand, they use the Google Maps API for implementation of the time-dependant vehicle routing problem (TDVRP), and on the other hand, they obtain historical travel time data from the Portland Transportation Achieved Listing (PORTAL).

Conrad and Fligiozzi used the Google Maps API (open-source) to obtain the optimised route taking into account high-quality network data, which incorporates road hierarchy and restrictions. By selecting the customer distribution and the depot on the screen, the interface calculates the shortest paths between customers and builds the distance and travel time OD matrices. Free-flow speeds are obtained using these travel times and distances from the Google Maps API. Using detailed traffic data from PORTAL (obtained from 436 inductive loop detectors), the speeds are later adjusted to take into account congestion effects, which will be applied at the bottleneck locations that can be added into the algorithm.

The proposed methodology developed by the authors to integrate real network data into the TDVRP was a significant improvement for representing the congestion effects in congested urban areas. However, the use of historical traffic data instead of real-time data limits their applications, especially in the context of the smart cities.

3.3.2   A Comparative Study of Vehicles' Routing Algorithms for Route Planning in Smart Cities [21]

Nha, Djahel and Murphy highlight the importance of considering dynamic road traffic conditions (congestion, incidents, etc.) when optimising routes within smart cities. The authors present a classification of different dynamic route planning algorithms and then compare their performance when they are applied in real networks.

To this end, first, the best route is calculated using the chosen algorithm (Dijkstra, Genetic Algorithm, etc.). Then, when a vehicle reaches a junction, the traffic conditions are checked and the route is re-calculated taking into account the updated traffic conditions. Finally, the algorithm stops when the final destination is reached.

To analyse the performance of this framework, the authors use the open-source traffic microsimulation software SUMO, together with the TRACI interface, which allows manipulating the simulation as it runs. As an example, they create a scenario with a specific road network in which the Dijkstra algorithm is applied. Then, the network is converted to SUMO format to simulate the initial route given by Dijkstra, and the simulation begins. Every time a vehicle reaches a node, the program checks whether there are changes in any road segment (link blocked by an accident or congestion), and if it is the case, the algorithm is re-applied ignoring those links.

To the best of the authors' knowledge, this paper is the first one in which a simulation software is linked to the optimisation algorithm to consider congestion effects. The implementation of this approach was in progress by the time the paper was written so there is a lack of details, for example, on how real traffic data is obtained, and how it is connected to the simulation software. In addition, this paper is not specifically focused on logistic activities. However, we think that this paper shows that it is possible to intertwine optimisation algorithms with simulation approaches leading to an increase in realism of the route planning processes, especially within smart cities.

3.3.3   Cyber-Physical Logistics System-Based Vehicle Routing Optimization [19]

The authors introduce the concept of cyber-physical logistics system (CPLS) which is a logistics system which incorporates sensors, heterogeneous communication network, and dynamic network interfacing as well as distributed computation technology to process the logistics information in real-time. Using real-time collection, transmission, and processing of the customers demand, vehicle status, and traffic information, the authors propose a routing adjustment model to minimise the total distribution cost. The decision-making centre collects all the information from customers (demands), road network (using traffic sensors) and delivery vehicles and optimise the routes.

Although the paper pays important attention to the communication network (TCP/IP, communication delays, data loss), the authors also focus on how road congestion is taking into account to optimise the delivery routes. They use a probabilistic model to calculate the congestion probability in the network. When a vehicle is travelling, the congestion probability should be lower than the setpoint. The decision-making process constantly update the congestion status and re-optimise the route if necessary. Therefore, the optimisation model (solved by a learnable evolution genetic algorithm) minimises the total distribution cost, constrained not only by customer time-window, vehicle maximum travelling distance, or vehicle capacity, but also by congestion.

To verify the proposed approach, the authors explain a case study in which they simulate a distribution system with 10 customer nodes, 1 distribution centre and 5 vehicles, and the results show the effectiveness of this method to reduce the total cost by re-adjusting the delivery routes taking into account congestion.

This paper is a good example on how new technologies and Intelligent Transport Systems can be applied to improve route optimisation algorithms. We think that this paper provides a good approach that could be useful in smart cities, and we strongly believe that future methodologies to develop VRP will be based on something similar to what is shown in this paper.

3.3.4   Solving the Dynamic Vehicle Routing Problem Under Traffic
Congestion [15]

In this paper, the authors propose a dynamic vehicle routing problem (DVRP) model with nonstationary stochastic travel times under traffic congestion. To this end, Kim et al. use a real case of a delivery company based in Singapore with a delivery network that consists of a single depot and multiple customers, where customers demand are known, but travel times between customer locations are time-dependent and stochastic due to traffic congestion. Real traffic information is also used.

Before estimating traffic congestion and travel time distribution, the authors highlight that the Euclidian distance (commonly used in other approaches) is inappropriate, and therefore they make use of the real road segments between two nodes. The traffic data were collected from the Land Transport Authority (LTA) in Singapore, which provides historical data from street sensors with which averaged vehicle speeds over time for different road segments can be calculated. When a delivery has to be made between two nodes, the Google Maps API is used to get the road segments that are part of the arch, and each road segment is linked to the historical speed data from the LTA (which is time-dependent). With the speed information available for each road segment, dynamics of the traffic congestion states and probabilities of each segment are estimated (see paper for details) and used as inputs to solve the dynamic vehicle routing problem through a rollout algorithm.

The comparison made in this paper between the results obtained through this approach and the current practice by the delivery company in Singapore (that ignores traffic congestion) highlights the potential saving from exploiting historical and real-time traffic congestion information, with a 7% improvement in total travel time. However, the paper is somehow similar to the one written by Conrad and Figliozzi, as they also used the Google API, and shares its limitations, as using historical data instead of real-time data (from Intelligent Transport Systems) can limit the applications of this approach.

3.3.5   Time-Dependent Vehicle Routing Problem with Path Flexibility [17]

In this paper, the authors have introduced the "Path Flexibility" concept, which consists on considering path selection in the road network as an integrated decision (according to the traffic congestion) in the time-dependent vehicle routing problem, minimising the total cost. The authors have applied this approach (TDVRP-PF) and compared it to the traditional one in the urban area of Beijing, using the speed patterns data of the road network from October 14th to 18th (2013). The roads have been classified depending on their average speed.

The results evidence that the TDVRP-PF may not only change the path between two customer nodes within the same route, but there can be structural changes in routes or even vehicle-customer assignments. Apart from this, two variants are included: TDVRP-PF-a (the assignments are fixed as in the TDVRP solution) and TDVRP-PF-r (the routes are fixed as in the TDVRP solution).

The proposed VRP modification developed by the authors to introduce path flexibility into the TDVRP allows minimising the total cost compared to the traditional VRP, making use of the congestion data to recalculate the routes using different paths. However, the use of traffic data from 4 days in 2013 introduces a sampling error in the data that could be fixed by using historical or real-time data.

3.3.6   Summary

Table 2 shows a summary of the papers studied. Some papers [15, 23] make use of the Google Maps API as the interface, and they use traffic data from official sources (databases) to analyse congestion effects. However, they use historical data rather than real-time data, which may limit their applicability. Others [19] develop a theoretical approach to be used within smart cities, in which traffic sensors, and in-vehicle technology can collect real-time information to gain insight on congestion costs that will be used in the route optimisation process, but the paper does not show any real case study or application. There is one paper [17] in which an approach is developed to take into account how congestion affects re-routing within urban areas, but the data used used is poor. And Nha et al. [21] propose the use of microsimulation software (using traffic congestion as an input) together with optimisation, but there is a lack of details on the overall methodology. It is also important to note that in most of the papers, congestion costs are considered through a probabilistic approach or based on averaged speeds, rather than estimating the real delay existing in each link of the network. Therefore, it can be seen that although each paper has its own advantages, there is a need to intertwine and improve the different approaches to further develop a more comprehensive and strong methodology.

Table 2. Summary of the main characteristics of the papers analysed.

	Type	Routing interface	Traffic data source	Traffic data type	Congestion calculation
Conrad and Figliozzi [23]	TDVRP	Google Maps API	Database. Inductive loops	Historical	Speed function
Nha et al. [21]	–	Microsim software	–	–	–
Lai et al. [19]	–	–	Sensors	Real-time	Probabilistic
Kim et al. [15]	DVRP	Google Maps API	Database. Inductive loops	Historical	Probabilistic
Huang et al. [17]	TDVRP-PF	–	Other	Weekly	–

4   Conclusions and Future Research

Traffic congestion is one of the main challenges within urban logistics. This paper has reflected how the research community has ignored this topic for decades whilst scholars were mainly focusing on the application of VRP at a strategic level, where minimising a distance-based cost function was the main objective. However, traffic congestion is dramatically increasing in our cities, and on-demand deliveries are changing the way we understand logistics activities. This has caused that, in parallel with the development of the so-called smart cities, congestion has raised concern amongst the research community in the last years.

However, there are only 30 Scopus-indexed papers in which the terms *"vehicle routing problem"*, *"logistics"*, and *"congestion"* have been considered together. In this paper, those articles have been analysed and rated taking into account the impact they are likely to have for other researchers. Then, the five papers with the highest scores have been described in order to understand how different authors account for congestion costs or driver behaviour within the VRP field.

The results show that there are two main approaches. The first one (more related to traffic engineering) makes use of traffic microsimulation software and intelligent transport systems (Big Data, sensors, communications technologies) to improve the realism of traditional approaches to consider real-time traffic congestion effects. However, these frameworks are very theoretical and may lack from direct applicability. On the contrary, there are some authors (more related to the operations research field) that have focused on the improvement of traditional approaches by considering traffic congestion as another input, using historical data and without considering the real congestion but an estimate given by a speed or probability function.

Considering congestion costs into route optimisation algorithms for logistics is something relatively new. Future VRP for urban logistics will have to consider congestion, driver behaviour effects and road conditions (accidents, road closures…) in real-time, making use of Intelligent Transport Systems and Big Data technologies. Agent-based traffic microsimulation or macrosimulation software can be also implemented as part of the optimisation approaches, to get realistic insights on how traffic conditions may affect routing and to analyse future and what-if scenarios.

To this end, the VRP needs to be addressed from a more comprehensive perspective. Urban logistics needs a multidimensional approach, and researchers from different fields (operations research, traffic engineering, smart cities…) need to play together if we want to improve how goods are delivered in the smart cities of tomorrow.

Acknowledgments. This work has been partially supported by the Spanish Ministry of Economy and Competitiveness (TRA2013-48180-C3-P and TRA2015-71883-REDT), FEDER, and the Ibero-American Program for Science and Technology for Development (CYTED2014-515RT0489).

References

1. Eksioglu, B., Vural, A., Reisman, A.: The vehicle routing problem: a taxonomic review. Comput. Ind. Eng. **57**(4), 1472–1483 (2009)
2. Dantzig, G., Fulkerson, R., Johnson, S.: Solution of a large-scale travelling salesman problem. Oper. Res. **2**, 393–410 (1954)
3. Dantzig, G., Ramser, J.H.: The truck dispatching problem. Manage. Sci. **6**(1), 80–91 (1959)
4. Clarke, G., Wright, J.: Scheduling of vehicles from a central depot to a number of delivery points. Oper. Res. **12**, 568–581 (1964)
5. Levin, A.: Scheduling and fleet routing models for transportation systems. Transp. Sci. **5**(3), 232 (1971)
6. Wilson, N., Sussman, J.: Implementation of computer algorithms for the dial-a-bus system. Bull. Oper. Res. Soc. Am. **19**(1) (1971)

7. Liebman, J.: Mathematical models for solid waste collection and disposal. In: 38th national meeting of the Operations Research Society of America Bulletin of the Operations Research Society of America, vol. 18, no. 2 (1970)
8. Golden, B., Magnanti, T., Nguyan, H.: Implementing vehicle routing algorithms. Networks 7(2), 113–148 (1972)
9. Juan, A.A., Faulin, J., Grasman, S.E., Rabe, M., Figueira, G.: A review of simheuristics: extending metaheuristics to deal with stochastic combinatorial optimization problems. Oper. Res. Perspect. 2, 62–72 (2015)
10. United Nations, World Urbanization Prospects: The 2014 Revision (2015)
11. Marks, D., Stricker, R.: Routing for public service vehicles. Locate Full-Text (Opens in a New Window) 97(UP2), 165–178 (1971)
12. Daganzo, C.F.: Distance travelled to visit N points with a maximum of C stops per vehicle: an analytical model an application. Transp. Sci. 18(4), 331–350 (1984)
13. Van Woensel, T., Kerbache, L., Peremans, H., Vandaele, N.: Vehicle routing with dynamic travel times: a queueing approach. Eur. J. Oper. Res. 186(3), 990–1007 (2008)
14. Srivatsa Srinivas, S., Gajanand, M.: Vehicle routing problem and driver behaviour: a review and framework for analysis. Transp. Rev., 1–22 (2016)
15. Kim, G., Ong, Y., Cheong, T., Tan, P.: Solving the dynamic vehicle routing problem under traffic congestion. IEEE Trans. Intell. Transp. Syst. 17(8), 2367–2380 (2016)
16. Lecluyse, C., Sorensen, K., Peremans, H.: A network-consistent time-dependent travel time layer for routing optimization problems. Eur. J. Oper. Res. 3(1), 395–413 (2013)
17. Huang, Y., Zhao, L., Van Woensel, T., Gross, J.: Time-dependent vehicle routing problem with path flexibility. Transp. Res. Part B Methodol. 95(1), 169–195 (2017)
18. Jabali, O., Van Woensel, T., De Kok, A.: Analysis of travel times and CO_2 emissions in time-dependent vehicle routing. Prod. Oper. Manage. 21(6), 1060–1074 (2012)
19. Lai, M., Yang, H., Yang, S., Zhao, J., Xu, J.: Cyber-physical logistics system-based vehicle routing optimization. J. Ind. Manage. Optimization 10(3), 701–715 (2014)
20. Cirovic, G., Pamucar, D., Bozanic, D.: Green logistic vehicle routing problem: routing light delivery vehicles in urban areas using a neuro-fuzzy model. Expert Syst. Appl. 41(9), 4245–4258 (2014)
21. Nha, V., Djahel, S., Murphy, J.: A comparative study of vehicles' routing algorithms for route planning in Smart Cities. In: 1st International Workshop on Vehicular Traffic Management for Smart Cities, VTM 2012 (2012)
22. Novaes, A., Bez, E., Burin, P., Aragao, D.: Dynamic milk-run OEM operations in over-congested traffic conditions. Comput. Ind. Eng. 88(11), 326–340 (2015)
23. Conrad, R., Figliozzi, M.: Algorithms to quantify impact of congestion on time-dependent real-world Urban Freight distribution networks. Transp. Res. Rec. 2168, 104–113 (2010)
24. Du, M., Yi, H.: Research on multi-objective emergency logistics vehicle routing problem under constraint conditions. J. Ind. Eng. Manage. 6(1), 258–266 (2013)
25. Ehmke, J., Steinert, A., Mattfeld, D.: Advanced routing for city logistics service providers based on time-dependent travel times. J. Comput. Sci. 3(4), 193–205 (2012)
26. Du, M., Yi, H.: Multi-objective emergency logistics vehicle routing problem: 'Road congestion', 'unilateralism time window'. In: 2nd International Conference on Logistics, Informatics and Service Science, Beijing (2013)
27. Polimeni, A., Vitetta, A.: Vehicle routing in urban areas: An optimal approach with cost function calibration. Transportmetrica B 2(1), 1–19 (2014)
28. Liang, X., Zhang, Y.: Study on vehicle routing problem with travel time coefficients. In: International Conference of Chinese Logistics and Transportation Professionals, Chengdu (2008)

29. Kok, A., Hans, E., Schutten, J.: Vehicle routing under time-dependent travel times: The impact of congestion avoidance. Comput. Oper. Res. **39**(5), 910–918 (2012)
30. Liang, Z.: Research of blocking factor combined with improved ant colony algorithm in VRP. In: 7th International Conference on Computational Intelligence and Security, Sanya (2011)
31. Xiao, J., Lu, B.: Vehicle routing problem with soft time windows. Adv. Intell. Soft Comput. **1**, 317–322 (2012)
32. Gupta, A., Heng, C., Ong, Y., Tan, P., Zhang, A.: A generic framework for multi-criteria decision support in eco-friendly urban logistics systems. Expert Syst. Appl. **71**(1), 288–300 (2017)
33. Muñoz-Villamizar, A., Montoya-Torres, J., Herazo-Padilla, N.: Mathematical programming modeling and resolution of the location-routing problem in urban logistics. Ingenieria y Universidad **18**(2), 271–289 (2014)
34. Soysal, M., Bloemhof-Ruwaard, J., Bektas, T.: The time-dependent two-echelon capacitated vehicle routing problem with environmental considerations. Int. J. Prod. Econ. **164**(1), 366–378 (2015)
35. Islam, S., Oslen, T.: Truck-sharing challenges for hinterland trucking companies: a case of the empty container truck trips problem. Bus. Process Manage. J. **20**(2), 290–334 (2014)
36. You, S., Chow, J., Ritchie, S.: Inverse vehicle routing for activity-based urban freight forecast modeling and city logistics. Transportmetrica A: Transp. Sci. **12**(7), 650–673 (2016)
37. Jin, X., Tang, Y., Xu, Q.: Routing optimization of city distribution considering access restriction. Appl. Mech. Mater. **505–506**, 959–966 (2014)
38. Danielis, R., Rotaris, L., Marcucci, E.: Urban freight policies and distribution channels. Eur. Transp. (Trasporti Europei) **46**, 114–146 (2010)
39. Zhu, X., Liu, T., Qiao, P.: The design and implementation of GIS logistics distribution system considering traffic information. Appl. Mech. Mater. **380–384**, 4671–4675 (2013)
40. Yin, Y., Liu, T., Tang, L., Li, Q.: Vehicle routing problem research based on information utility theory. Gummi, Fasern, Kunststoffe **69**(14), 2084–2090 (2016)
41. Figliozzi, M.: The impacts of congestion on time-definitive urban freight distribution networks CO_2 emission levels: results from a case study in Portland, Oregon. Transp. Res. Part C Emerg. Technol. **19**(5), 766–778 (2011)

Distributed Genetic Algorithms on Portable Devices for Smart Cities

J.A. Morell[(⊠)] and Enrique Alba

Departamento de Lenguajes y Ciencias de la Computación,
University of Málaga, Andalucía Tech., Málaga, Spain
{jamorell,eat}@lcc.uma.es

Abstract. In the future smart city, citizens are interconnected and easily share information anywhere, anytime. Through a sensor network integrated with real time monitoring systems, data are collected, processed and analyzed. Of course, this is already happening, in part. Nowdays, the number of portable devices that are available to all people is huge and we can find them everywhere, they are not only smartphones but also tablets, IoT, and other. This is a perfect scenario to start new lines of research on the actual suitability of portable devices to solve real optimization and machine learning problems. On the one hand, the fact that they are everywhere encourages research aimed at their collaboration in a distributed way. On the other hand, genetic algorithms are metaheuristics where parallelization takes on great importance. In this paper, we analyze the numerical behavior of distributed genetic algorithms on portable devices. We focus on the behavior of the distributed algorithm when we modify the number of interconnected devices, as well as the behavior of the algorithm when the devices with different performances collaborate together. As a conclusion, the numerical results support the future research in the concept of distributed intelligence everywhere, since algorithms worked out accurate and efficient results.

1 Introduction

During these last years, portable devices of all kinds are reaching homes, businesses, and cities. Not just mobile phones and tablets are ubiquitous, but also smartwatches, Raspberry Pi's, and other small computers that perform tasks everywhere.

Nowdays, one of the most important lines of research in the field of metaheuristics is the one that deals with the analysis of the behavior of complex algorithms [1] in non-traditional environments. Given that genetic algorithms [2,3] are eminently parallel [4], and given the large number of portable devices everywhere, it seems clear that their combination is a research idea that should be studied.

This research has been partially funded by I Plan Propio de Investigacón y Transferencia de la Universidad de Málaga 2016–2017. Also, partially funded by the Spanish MINECO project TIN2014-57341-R (http://moveon.lcc.uma.es).

© Springer International Publishing AG 2017
E. Alba et al. (Eds.): Smart-CT 2017, LNCS 10268, pp. 51–62, 2017.
DOI: 10.1007/978-3-319-59513-9_6

In this article, we present a comprehensive analysis, design, implementation and evaluation of genetic algorithms running on portable devices such a smartphones and tablets. We analyze what is the behavior of these algorithms, as well as we compare the results of the distributed one to canonical GAs when computing smart routes for users.

For this analysis, a problem of the real world has been chosen, what will give an additional value to our study. The problem chosen for the experimentation is the Capacitated Vehicle Routing Problem (CVRP) [5,6] which is explained in detail in Sect. 2.

Our main aim is to analyze the numerical behavior of distributed genetic algorithms on portable devices to understand how to proceed for larger and wider works in the future. We study the adaptation of these devices and the resulting behavior of the algorithm when the number of interconnected devices are dynamically modified in a distributed environment: a very real scenario.

In summary, we will answer the following set of worth research questions:

- RQ1: Are portable devices suitable to solve complex problems using GAs?
- RQ2: Can we use the same traditional algorithms in these devices, or do we have to adapt them?
- RQ3: Can we implement a distributed genetic algorithm and make it substantially better than a canonical GA in this kind of devices?
- RQ4: How does the algorithm behave when putting together devices with different performances?

This article is organized as follows: Sect. 2 presents the chosen problem. Section 3 describes the algorithm we use to solve the chosen problem. Section 4 shows details about the implementation. Section 5 analyzes and discusses the experiments that we do and the results that we obtain. Finally, Sect. 6 draws the main conclusions and the future work research.

2   The Capacitated Vehicle Routing Problem

For the experimentation, we will use a real problem (CVRP) [7] that will let us analyze the behavior of an intelligent algorithm in a distributed environment on portable devices. This is indeed a potential final real application, since routing in a city appears in many forms in the new challenges of *smart mobility*.

VRP consists in delivering goods to a set of customers with known demands through minimum-cost vehicle routes originating and terminating at a depot. VRP is a well-known integer programming problem which falls into the category of NP-hard problems [8], needing advanced non-exhaustive techniques for its resolution. Solving VRP is equivalent to solving multiple TSPs at once.

There are many extensions of VRP. We have considered in this paper the Capacitated Vehicle Routing Problem (CVRP) [9], in which a fixed fleet of delivery vehicles of the same capacity must service known customer demands for a single commodity from a common depot at minimum total transit costs [10,11].

3   Our Approach to Solve CVRP with a GA

GAs are inspired in biological evolution. They are especially useful to solve NP-hard and even NP-complete problems. There is a wide variety of problems in which we need to maximize or minimize a certain function. On the one hand, exact methods allow exact solutions to be found; however it is difficult to solve NP-hard problems with exact methods when the size of the problem increases. On the other hand, metaheuristics provide suboptimal (maybe optimal) solutions in a reasonable time. Hence, GAs come handy in these situations.

Algorithm 1. Pseudocode of the Genetic Algorithm (GA)

$totalEvaluations \leftarrow 1$
$population[] \leftarrow Initialize\_Population()$
for $i < population.size$ **do**
    $population[i] \leftarrow Evaluate\_Indiv(population[i])$
$population \leftarrow Compute\_Stats(population)$
while !Is_Finish() **do**
    $parents[2] \leftarrow Select\_Individuals2(population)$
    $offspring \leftarrow Recombination(parents[0], parents[1])$ //According to Pc
    $offspring \leftarrow Mutation(offspring)$ //According to Pm
    $offspring \leftarrow LocalSearch(offspring, population)$ //According to Pls
    $offspring \leftarrow Evaluate\_Indiv(offspring)$
    $population \leftarrow Replace\_Worst(offspring, population)$
    $population \leftarrow Compute\_Stats(population)$
    $totalEvaluations + +$
$solution \leftarrow Get\_Better\_Individual(population)$

GAs must be customized for each problem. In this section, we explain our first approach to solve a CVRP with a basic GA (Algorithm 1). Our ultimate goal is to design a distributed GA for portable devices, but we have to get first a basic good algorithm to start with.

To this end, we use specific variation operators and a local post-optimization as an added stage. We have implemented our algorithm similarly to that used by [10,11].

A solution is represented by a vector, containing a permutation of customers and route splitters. Two contiguous route splitters mean an empty route. The beginning and the end of the vector act as a virtual route splitter [10,11].

We use binary tournament selection, which consists of randomly selecting two different individuals from the population and then choosing the one with better fitness between them. Then the process is repeated once again, obtaining a new individual that must be different from the first one.

For recombination, we will use the Edge Recombination Crossover (ERX) [11]. This is a crossover technique for permutation chromosomes. ERX focuses in preserving the links between customers (edges of the implicit graph).

As in nature, mutation is useful, introducing a degree of diversity needed in each generation. In spite of the result of mutation (positive or negative), it allows to explore a larger state spaces helping to find a better solution over time. We used three mutation operators [11] shown to be appropriate for CVRP: Insertion, Swap, and Inversion. All of them have an equal probability to operate on every gene. They are used in a combined way. Anytime a mutation is going to happen to a gene, one of these operators is selected with equal probability.

According to the literature [11], in addition to the classical operators (recombination and mutation), it is almost mandatory to use a local search method to obtain results of high quality in the CVRP. We have added a local refining step consisting in applying *2-Opt* and *1-Interchange* [11], one after another in that order. Local search takes place after recombination and mutation operators with an associated probability.

For computing the fitness of a set of routes (a solution) S, we use an existing fitness function [11]: $f(S) = F_{CVRP}(S) + \lambda \cdot overcap(S) + \mu \cdot overtm(S)$ [10].

The aim is to minimize the fitness. The $overcap(S)$ function is equal to the excess items over the maximum allowed capacity per truck. The $overtm(S)$ function is equal to the excess hours over the maximum allowed per truck. If no excess exists in the tentative solution S then penalization is equal to zero in both cases.

We multiply $overcap(S)$ and $overtm(S)$ by constants to weight the result $\lambda = \mu = 1000$. This is done to penalize those infeasible solutions that do not meet the restrictions: they will have less chance of survival.

4   Design and Implementation of a Distributed GA for Portable Devices

In this section, we show some details about the design and implementation of a distributed GA where multiple portable devices running the previous algorithm interact with each other. Details are many, since running optimization techniques in smartphones is not a well-known discipline [12], so we will try to summarize here most of the design and implementation of the algorithm on portable devices (Android, iOS, Raspberry PI, laptops, etc.). First, we identify the actors. Second, we explain the layered design of the application. Finally, we emphasize the robustness of the communications in our model.

4.1   Actors Identification

We have designed an architecture to develop a prototype that allows us to approach the problem in our first attempt. We use a star topology.

The application is run from an arbitrary portable device by opening a web browser from where the user configures the problem parameters and starts the problem (see Fig. 3). This is a cross-platform architecture that allows any type of portable device to join in and leave out at will.

There are two actors involved in the app. The slave process, which is executed in each portable device, and the monitor process, which is executed in the server.

There are as many slave processes as portable devices. There is only one monitor process running on the server (desktop computer).

In this approach, the entire the population is copied in all slaves and the monitor. Furthermore, the monitor does not generate new individuals; it is responsible for maintaining the population updated by remplacing the worst individuals of the population by the new individuals received from the slave processes. We have to think of the monitor in a passive way, specialized in guiding the search with replacements, not in the creation of new solutions.

The monitor process only has to replace the worst individual in the population by the individual received by the slave process, keeping the population as up-to-date as possible. When the monitor receives a new individual from a slave, it responds to the slave with a message requesting it to generate a new individual. Also, next to this message, the monitor sends the last changes in the population to the slave. Only when the new individual is better (or equal) than the worst individual in the population replacement is made. In this way, it is not always necessary for the monitor to send changes to the slaves.

The reason for not creating new individuals in the monitor process is because the monitor must be prepared to receive messages from a large number of slave processes. That is, if there are already hundreds of slave processes creating individuals, the fact that the monitor also generates individuals does not make any difference and would delay its work. In addition, since the monitor is the only one with the most up-to-date population, it is common sense that it is responsible for replacing the worst individual in the population with the new individual created by the slave process. By specializing the monitor process in the above tasks, we make it more efficient. If we add the need to create new individuals and compute their fitness we can overload the monitor too much and not be efficient in the other tasks, that are more important. In addition, we are interested in keeping a server as light as possible, as processing on server is expensive while processing on slaves is free.

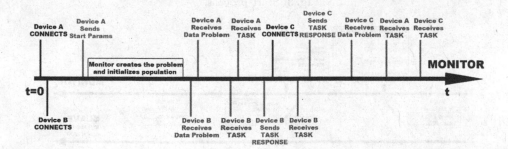

Fig. 1. Example of timeline.

As to slave devices, each one receives orders to create new individuals from the monitor. The slave creates new individuals and send them to the monitor. However, the slave needs to update its population before creating any individual.

For this purpose, the monitor maintains a structure to store all changes in population relative to the population of each slave process. This structure consists of an array of bits for each slave in which each position is marked as 1 if that individual has changed, and as 0 if it has not changed. Therefore, if there are 1000 slaves and a population of 1000 individuals, this structure has a size of 1 Kb for each slave, and a total size of 1 Mb (125 KB). The monitor sends the changes in the population to the slave process in a message in which the monitor orders the slave to generate a new individual (see Fig. 4).

We could think that this architecture (see Fig. 1) is not scalable or that too many messages are sent. However, we seek to solve problems where fitness

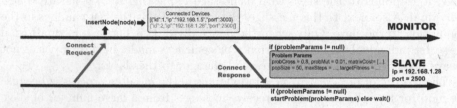

Fig. 2. Connection process.

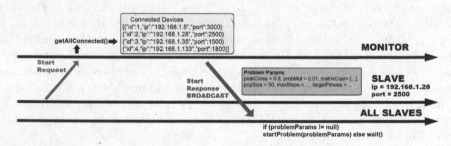

Fig. 3. Start process.

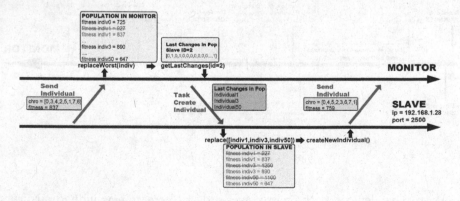

Fig. 4. Task process.

computation is the slowest process. Sometimes, it is necessary to carry out fitness computations that are slow, or that require user interaction, or in which the use of sensors to interact with the environment is necessary to compute fitness. These are the types of problems in which this architecture excels. Thanks to modern mobile phones, we have a sensor network moving through the city that can be used to analyze the environment and draw conclusions at the same time about the quality of the fitness found. In addition, smart cities will not only have mobile phones but we will also have small computers everywhere that can be used for this purpose.

Moreover, in this architecture, users can enter and exit the problem at any time while the algorithm continues to run without losing information. We can solve problems in a very dynamic environment that has to be prepared so that hundreds or thousands of users can collaborate. When a new slave connects to the server during the problem resolution, the algorithm data (Sect. 4.3) and the current population are sent to the slave so that it can join to collaborate even if the rest of devices have been running for a long time before its arrival (Fig. 2). If one of the slaves disconnects, the monitor deletes its data and continues to solve the problem with the other slaves without any issue. When a user disconnects from the server, there is no loss of information because the server has the population updated and the slave device only has a copy of it.

If we had used a traditional island architecture, there is the problem that each island must generate a new population when it connects. We have to decide how to create this new population. We must also prevent losing information when a user disconnects. We have to think how to save the best individuals of that user. We can not forget that all this happens in a dynamic and unpredictable environment, so it is difficult to take into account all these variables. However, this is a study that we have to do in the future. This is our first architecture where we foster the dynamic availability of devices versus memory efficiency: redundancy is a low price in a dynamic platform of arbitrary users.

4.2   Distributed Application Layers

We divide the distributed application into three modular layers with different functions: transmission layer, communication layer and application layer.

First, the transmission layer encapsulates the communication protocol used. In this way, we can easily change the communication protocol (UDP, TCP, Web-Sockets, etc.). In this work we are using WebSockets.

Second, the communication layer encapsulates the operations of sending and receiving messages. When the monitor sends a message to a slave with a task, it changes the status of that node to *busy* until it receives the response. Thereby, the other node is allowed to complete the task before receiving a new one. An order to create a new individual is a task. We plan to expand the types of tasks in the future. Moreover, the communication layer maintains a list of available nodes. A node is another device to communicate with. Finally, the application layer contains the logic of the application.

4.3   Robustness of Communications

The layered design allows the application layer to be totally independent of the number of connected devices. This design recovers from failures in the connection during the running of the distributed algorithm, and allows a dynamical number of portable devices joining or leaving the processing pool at will. In this first work we keep constant the number of portable devices during all the runs, but of course make many different runs with different number of devices in them.

Each time the monitor sends a message, the application layer asks the communication layer for the ID of a free node. hat a node is free means that the previous task has already been completed and the node is available again.

When a slave completes a task, it sends a response message to the monitor with the created individual. Next, when the monitor receives a response message from a slave it marks that slave as free. As the monitor receives responses from the slaves who had assigned tasks, it sends new tasks to the free slaves. When the monitor sends a new task to a slave, it also sends the last population changes.

5   Experimentation

This section presents the experimental settings used to assess the performance of distributed genetic algorithms on portable devices. Later, the experiments that have been carried out and the results obtained are also presented.

We have designed and built the implementation of *UbiquitousJS*, a multiplatform JavaScript library of distributed metaheuristic algorithms for solving optimization problems. *UbiquitousJS* focuses on portable devices and it has developed by us for this project.

For experimentation, 4 different portable devices have been used: Lenovo Tablet {a}, Neo Mobile {b}, Lenovo Mobile {c}, Samsung Tablet {d}, Amazon EC2 Instance (*monitor*). As monitor we use an Amazon EC2 instance whose function is to keep the population updated and store the list of replacements. As portable devices we use 2 mobile phones and 2 tablets of different specifications.

The instance chosen for CVRP experimentation is CMT6 [13] which has a size of 50 cities and a depot. The selected parameters are: recombination probability 0.8, mutation probability 0.01, local search probability 0.1, vehicles number 6, capacity penalty 1000, time penalty 1000 and target fitness 650. All parameters relating to the problem setting have been set to good previous values [11]. We use a copy of the population of 20 individuals on each device and on the monitor. For each experimental instance, we have carried out 30 independent runs of our algorithm. A different random seed was selected for each repetition.

With this experimentation we want to analyze the performance of the distributed algorithm when 4 portable devices collaborate with each other. To measure the performance of the distributed algorithm we use their speedup. However, we need to compare the times obtained when we use 4 devices and when we use only 1 device to observe if there is an improvement in performance. But, what is the device we have to choose to compare to? We know that each device has different specifications, so each will have a different performance. Therefore, firstly

Table 1. Experiment results. Target fitness = 650.

ids	Time (ms)				Evaluations		
	Avg ± stdv	Max	Min	Friedman rank	Avg ± stdv	Max	Min
{a,b,c,d}	**30612 ± 8990**	62190	**16255**	**1.83**	1251 ± 380	2702	692
{a,b,c}	32683 ± 4781	**43303**	22611	2.00	1109 ± 162	1474	814
{a,b}	43908 ± 11047	71269	28249	3.57	1048 ± 244	1641	707
{a,d}	44801 ± 9440	65320	28070	4.00	996 ± 196	1519	653
{d,c,b}	47521 ± 10723	75747	33400	4.10	1175 ± 260	1887	873
{d,c}	61860 ± 12393	83380	38230	5.77	1004 ± 211	1366	545
{a}	79740 ± 16573	122728	59175	6.77	1007 ± 211	1584	725
{b}	97095 ± 3702	108997	98403	7.97	1007 ± 211	1584	725
{c}	114710 ± 7277	121734	98757	9.07	1007 ± 211	1584	725
{d}	132903 ± 3652	118939	101511	9.93	1007 ± 211	1584	725

we have to measure the different performances of all of the portable devices to know what we are comparing to. This is our first set of experiments.

As a result of this set of experiments we got the times of the 4 devices sorted from best to worst time. Since we deal with stochastic algorithms, we have performed a Friedman statistical test to obtain the ordered results in a ranked list (see Table 1). Device $\{a\}$ 80s, device $\{b\}$ 97s, device $\{c\}$ 115s and device $\{d\}$ 133s.

Once we have the times of each of the portable devices, it is interesting to measure different combinations of collaboration between them. It could happen that a combination with fewer portable devices achieves lower times. Nevertheless, we do not need to test all combinations. We are interested in analyzing whether there is an improvement in performance if we repeat the experiment by adding a new device of better performance (or worse performance) than those that were available already.

Our second set of experiments consists of a single experiment in which we analyze the performance of the distributed algorithm when device $\{a\}$ and device $\{d\}$ collaborate with each other ($\{a,d\}$). Both are the devices that got the best and the worst times in the first set of experiments. We want to see if there is an improvement in performance against the best of them when we add additional (slower) devices (Fig. 5).

We have obtained an improvement in avgt of 1.78 times from $\{a\}$ to $\{a,d\}$ and an improvement in avgt of 2.97 times from $\{d\}$ to $\{a,d\}$. These are very good results, since in both cases there has been an improvement in performance. Even in one case this improvement has been super-linear.

Our third set of experiments is composed of three experiments. The first one is performed with the two devices that obtained the worst times in the first set of experiments $\{c,d\}$. Then the experiment is repeated with the three devices that

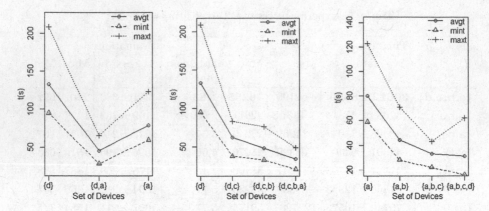

Fig. 5. Second set of experiments.

Fig. 6. Third set of experiments.

Fig. 7. Fourth set of experiments.

obtained the worst times $\{b, c, d\}$. Finally, the experiment is repeated with the all devices $\{a, b, c, d\}$. We want to see whether the performance improvement is linear or not, and whether performance improvement decreases as devices increase when we repeat the experiment by adding a new device of better performance than those that were already.

As we can see in Fig. 6, there is a super-linear speedup when we repeat the experiment by adding a new device of better performance. We have obtained an improvement in avgt of 2.15 times from $\{d\}$ to $\{c, d\}$, an improvement in avgt of 2.80 times from $\{d\}$ to $\{b, c, d\}$, and an improvement in avgt of 4.34 times from $\{d\}$ to $\{a, b, c, d\}$. One might think that this is expected because if we add devices with better performance the speedup should be super-linear. However, this is not so common, and it is interesting to show that this is true in practice.

Our fourth set of experiments is similar to the third, but in reverse order. It is composed of two experiments. First, an experiment is performed with the two devices that obtained the best times in the first set of experiments. Then the experiment is repeated with the three devices that obtained the best times. We do not need to repeat the experiment with all the devices together because it has already been done before. We want to see whether the performance improvement is linear or not, and whether performance improvement decreases as devices increase when we repeat the experiment by adding a new device of worse performance than those that were already.

As we can see in Fig. 7, there is a sub-linear speedup when we repeat the experiment by adding a new device of worse performance. We have obtained an improvement in avgt of 1.82 times from $\{a\}$ to $\{a, b\}$, an improvement in avgt of 2.43 times from $\{a\}$ to $\{a, b, c\}$, and an improvement in avgt of 2.60 times from $\{a\}$ to $\{a, b, c, d\}$.

These results are as expected, because if we add devices with worse performance the speedup should be sub-linear. Nevertheless, we have to notice that there is an improvement anyway. Although the slower devices do not help much, they are able to partially improve the speedup results. This is promising and invites us to carry out new experiments with a greater number of devices in the future in order to analyze how many devices would be able to support this architecture without collapsing.

6 Conclusions

In this article we propose a distributed genetic algorithm focused on portable devices. To achieve our goal, we have designed a modular, layered architecture and we have implemented it in JavaScript (*UbiquitousJS*) to allow an easy and fast connection to the set of voluntary collaborators by a simple web browser (RQ2). Distributed devices communicate with each other via the Internet. This algorithm allows users to collaborate in solving a wide variety of problems in smart cities.

In this study, we wanted to make a first approximation to this problem. In this first approach, we have designed an architecture that has served as a prototype for experimentation. However, we know that we are still far from having an architecture that can be used in real environments. This is something we have pending for future works.

Nevertheless, the results obtained in the experimentation have been encouraging. We have shown how several portable devices (RQ1) of different specifications (RQ4) collaborating together have been able to improve their performance separately (RQ3). In addition, we have designed an architecture that allows users to join the collaboration at any time and disconnect whenever they want, without the algorithm losing information. This is perfect for dynamic and unpredictable environments such as smart cities. This is what leads us to believe that we are on the right track.

Of course, there are still many challenges that need to be addressed. It is necessary to continue this line of research to analyze what is the maximum number of devices allowed by this design until saturation occurs. In addition, we have to experiment with heterogeneous devices to try to reduce the disadvantages that occur when slow devices collaborate with fast devices [14].

Moreover, the fact that the connected nodes can vary is interesting for future research, because it allows the resolution of problems without making use of dedicated devices. Furthermore, new devices only have to access a web page to connect to the monitor, so the connection process is really easy as well. In future work we must analyze what happens when new nodes connect and disconnect while the algorithm is running.

In short, a deeper investigation is still needed to develop an architecture that allows us to solve real complex problems of smart cities by distributed genetic algorithms. However, the first results are encouraging and now we have a baseline for numerical comparisons in the future.

References

1. Alba, E., Blum, C., Asasi, P., Leon, C., Gomez, J.A.: Optimization Techniques for Solving Complex Problems, vol. 76. Wiley, Hoboken (2009)
2. Bäck, T., Fogel, D.B., Michalewicz, Z.: Handbook of Evolutionary Computation. Oxford, New York (1997)
3. Michaelwicz, Z.: Genetic Algorithms + Data Structures = Evolution Programs. Springer, Heidelberg (1992)
4. Alba, E., Tomassini, M.: Parallelism and evolutionary algorithms. IEEE Trans. Evol. Comput. **6**(5), 443–462 (2002)
5. Letchford, A.N., Lysgaard, J., Eglese, R.W.: A branch-and-cut algorithm for the capacitated open vehicle routing problem. J. Oper. Res. Soc. **58**(12), 1642–1651 (2007)
6. Wassan, N.A., Wassan, A.H., Nagy, G.: A reactive tabu search algorithm for the vehicle routing problem with simultaneous pickups and deliveries. J. Comb. Optim. **15**(4), 368–386 (2008)
7. Dantzig, G.B., Ramser, J.H.: The truck dispatching problem. Manage. Sci. **6**(1), 80–91 (1959)
8. Gary, M.R., Johnson, D.S.: Computers and intractability: a guide to the theory of NP-completeness (1979)
9. Alba, E., Dorronsoro, B.: Solving the vehicle routing problem by using cellular genetic algorithms. In: Gottlieb, J., Raidl, G.R. (eds.) EvoCOP 2004. LNCS, vol. 3004, pp. 11–20. Springer, Heidelberg (2004). doi:10.1007/978-3-540-24652-7_2
10. Alba, E., Dorronsoro, B.: Cellular Genetic Algorithms, vol. 42. Springer Science & Business Media, New York (2009)
11. Alba, E., Dorronsoro, B.: Computing nine new best-so-far solutions for capacitated vrp with a cellular genetic algorithm. Inf. Process. Lett. **98**(6), 225–230 (2006)
12. Cintrano, C., Alba, E.: Genetic algorithms running into portable devices: a first approach. In: Luaces, O., Gámez, J.A., Barrenechea, E., Troncoso, A., Galar, M., Quintián, H., Corchado, E. (eds.) CAEPIA 2016. LNCS, vol. 9868, pp. 383–393. Springer, Cham (2016). doi:10.1007/978-3-319-44636-3_36
13. Christofides, N.: Combinatorial optimization. In: Nicos, C. (ed.) A Wiley-Interscience Publication, Based on a series of lectures, given at the Summer School in Combinatorial Optimization, held in Sogesta, Italy, 30 May - 11 June 1977. Wiley, Chichester (1979)
14. Nesmachnow, S., Cancela, H., Alba, E.: Heterogeneous computing scheduling with evolutionary algorithms. Soft. Comput. **15**(4), 685–701 (2010)

Existing Approaches to Smart Parking: An Overview

Fernando Enríquez[1], Luis Miguel Soria[1]([⊠]), Juan Antonio Álvarez-García[1],
Francisco Velasco[2], and Oscar Déniz[3]

[1] Computer Languages and Systems Department, University of Seville, 41012 Seville, Spain
{fenros,lsoria,jaalvarez}@us.es
[2] Applied Economics I Department, University of Seville, 41018 Seville, Spain
velasco@us.es
[3] VISILAB, E.T.S.I.I, University of Castilla-La Mancha, Ciudad Real, Spain
oscar.deniz@uclm.es

Abstract. After years of technological advances, parking is still a problem for many people. It is a time-consuming task that we all have to face on a day-by-day basis and it is also a problem for cities, that see how traffic and pollution increases. There have been multiple attempts to find a partial or global technological solution to this problem, ranging from using different types of sensors or cameras for automatically detecting free spaces to collaborative apps that let users share relevant information. In this paper, we give an overview of the methods developed so far, showing their main features, differences, pros, and cons, as well as other factors that may contribute to the success or failure of new proposals that will come in the future.

Keywords: Smart city · Parking · Crowdsensing · Computer vision

1 Introduction

According to the United Nations [1], 54% of the population lives in urban zones and this is expected to grow to 66% in 2050. This tendency makes urban mobility more difficult and makes finding a parking space one of the most repetitive problems for citizens in big cities. Furthermore, vehicles cruising for parking are responsible for at least 30% of traffic jams [2], with the average time for search being more than 20 min [3]. Besides the personal problems that this can generate, this is a problem in terms of fuel consumption, CO_2 emissions and in general a waste of resources for the community.

Beyond private spaces, there is no worldwide accepted solution for monitoring vehicles that enter or leave an area, though several efforts have been made to solve the problem in a specific context. *Smart parking* is the term used for a set of technologies and applications targeting issues related to parking in Smart Cities and this work aims at providing an up-to-date survey of the most interesting and relevant solutions. The rest of the paper is organized as follows. Section 2 analyses the state of the art in Smart Parking solutions. The main conclusions are drawn in Sect. 3.

© Springer International Publishing AG 2017
E. Alba et al. (Eds.): Smart-CT 2017, LNCS 10268, pp. 63–74, 2017.
DOI: 10.1007/978-3-319-59513-9_7

2   Classification

Parking solutions are offered off-street or on-street. Off-street parking may refer to mul-tistory car parks, while on-street refers to parking spaces along public roads and streets. This work considers both types of scenarios. Typically there is a focus on infrastructure, where several sensors are installed to monitor the places, crowdsensing where the drivers' smartphones are the source of the information or a hybrid proposal. Here, vision solutions are analyzed as a specific category due to the increase of cameras and the potential of this type of sensor.

2.1   Infrastructure

In order to get the occupancy status of parking places, fixed or mobile sensors are installed on/off-street to detect vehicular events. Mobile sensors are not the most com-mon solution although Parknet [4] in San Francisco is a prominent study: taxi cars col-lected the occupancy status of the parking place when they passed beside it, gathering data from GPS receiver and ultrasonic sensors. Although every taxi can detect multiple spots, updating the information of the same spot can take 25 min with a fleet of 300 vehicles. Parking Spotter is based on the same idea, leveraging sonar and radar sensors of some Ford vehicles. There are also some LiDAR-based solutions [5] although only focused on surveying parking spaces with one equipped car. These works are hardly scalable since all the vehicles must have the same kind of sensor or else share the infor-mation through the same database and the number of mounted sensor-vehicles must be enough to update the information frequently.

Fixed sensors are the more extended and popular option. Although there are a wide variety of sensors [6] (active infrared, ultrasound, acoustic, accelerometer, etc.), magnetometer is by far the most common fixed sensor. The magnetometer is accurate although normally solutions require at least one sensor per place, increasing the cost of the deployment. The sensor itself measures the current magnetic fields and detects the presence of metal vehicles. Most municipal deployment projects or large shopping centers drill magnetometers in parking places, sharing this information through visual signals or mobile applications. Installation and maintenance processes involve access to the property and road surface so it is only possible for off-street or on-street with government permission.

Table 1 refers to the most prominent smart city parking solutions. SmartSantander and San Francisco can be seen as pilot studies. SmartSantander was conceived as a Smart City Laboratory and San Francisco finished its pilot study at the end of 2013 due to the cost of sensor maintenance. Nice dropped their mobile application Nice Passport in 2016 after some organizational[1] and security problems[2]. Málaga, London (the only one based on RFID), Moscow or Los Angeles are success cases of Smart Parking.

---

[1] http://www.20minutes.fr/nice/1839579-20160504-nice-trois-ans-apres-installation-stationne-ment-intelligent-disparait.

[2] http://www.lefigaro.fr/politique/le-scan/couacs/2014/06/06/25005-20140606ARTFIG00112-securite-informatique-a-nice-pris-de-court-estrosi-interrompt-une-interview.php.

Table 1. Infrastructure parking projects

Smart City	#places	Company	URL	Year
San Francisco[a]	8.2K	Fybrtech	http://www.fybr-tech.com	2011
Santander [8]	0.4K	Libelium	http://bit.ly/2mOd38r	2011
Nice[b]	4.5K	Urbiotica	http://www.urbiotica.com	2012
Los Angeles [9]	6.3K	StreetLine	http://lat.ms/1BVDxpD	2012
London[c]	3.4K	Smart Parking	http://bit.ly/2mTSTcm	2012
Moscow[d]	50K	WorldSensing	http://bit.ly/2lCMAbU	2012
Malaga[e]	2.2K	Parkhelp	http://bit.ly/2mTPNFc	2014

[a] http://sfpark.org/resources/parking-sensor-technology-performance-evaluation
[b] http://www.urbiotica.com/en/inauguration-of-our-smart-parking-project-in-the-city-of-nice-fr/
[c] https://www.westminster.gov.uk/parkright
[d] http://parking.mos.ru/en
[e] http://www.eesc.europa.eu/?i=portal.en.events-and-activities-smart-cities-malaga

Research for fixed sensor-based solutions is focused on reducing the installation time and cost [7] using surface-mounted magnetometers that can be glued to the road and enlarging battery life of wireless sensors.

2.2 Vision

Due to the increasing interest of the scientific community and users in general in artificial intelligence techniques based on images, in recent years more solutions are appearing using artificial vision. These systems, unlike previous infrastructures, are not yet established, being mostly used in controlled and experimental environments. Although there are companies that base their systems on these techniques, most of the work is still under development.

In general, the detection of parking spaces using vision is subject to several problems. The first one is the quality of the image. For certain works based on object recognition, the image must be of sufficient quality to be processed. During the day, this may not be an obstacle. However, in conditions of insufficient lighting or adverse weather conditions, it becomes a real problem. Another problem are occlusions. Depending on the location of the image capture system, the vehicles themselves or surrounding elements (trees, buildings, street furniture, shadows) may obstruct the view of the parking area. This problem can be solved by changing the placement of the capture system, although this may not always be possible, or reduce the monitored surface due to the change in perspective. Finally, an inherent problem with this detection technique is classification itself. Image-based classification systems have proliferated over the past decade, although they are still far from offering the assurance of systems based on structural elements.

On the other hand, vision algorithms have advantages in terms of coverage, cost, and versatility. Since a single capture system can cover dozens (or hundreds) of parking spaces, the cost reduction compared to infrastructure-based systems is significant.

In addition, the maintenance cost of these systems is negligible, except in those cases where drones or satellites are required to acquire the images. The latter are usually multi-functional, which leads to another strength: versatility. Since the images obtained can be used not only for parking control, systems implementing the vision-based solution are often used for other purposes at the same time. Examples of this functionality are surveillance systems, pedestrian control, maintenance tasks, and other scenarios that increase the return of the investment.

Solutions based on vision can use different systems to get the images of the zone being monitored. Among other, the most common methods for the image acquisition are external video cameras, vehicles equipped with vision systems, three-dimensional capture systems, and zenithal or aerial images obtained by satellites or drones.

In the same way, as discussed above, detection systems are supported by a set of vision algorithms. These algorithms can be grouped according to the analysis performed on the images. Most works in the state of the art base their development on the following approaches:

- Appearance based approaches. Based on the comparison of the current appearance of a parking place with an original appearance of the vacant state. Many techniques are tailored and fine-tuned to specific contexts and scenarios. However, these techniques can not be easily generalized, and even the adaptation of one solution to a different parking lot is not straightforward.
- Recognition based approaches. Approaches based on object recognition aim to detect and classify the vehicles occupying the parking space using machine learning algorithms. This is a complex approach because of the large variety of the target objects.
- Three dimensional image processing.
- Combined techniques applying image processing in order to improve the quality and avoid light variation effects, and machine learning algorithms to classify image content.

Finally, monitoring techniques can be divided into two types depending on the way in which lots are processed:

- Estimating occupancy of an entire parking lot, for example, by counting incoming vehicles.
- Checking for the presence of a vehicle in each cell. Most vision-based approaches require the presence of vehicles in individual parking lots.

Below, a compilation of some relevant papers from the related bibliography are shown. Among them, there are different approaches as a sample of the heterogeneity of the existing processing, recognition, and image acquisition systems.

In [10] a distributed and efficient system is proposed to solve the problem of parking with vision systems. For this purpose, convolutional neural networks specifically designed for smart cameras are used. Two visual datasets have been used to verify the accuracy of the proposed system: PKLot and CNRPark-EXT (this latter dataset was created by the authors). Thanks to convolutional neural networks applied in the classification process, the proposed solution is robust to images exposed to partial occlusion, shadows, and changes in light conditions. In addition, it has a good capacity of generalization. A reduced version of the AlexNet neural network was used in this paper. The new convolutional neural network is able to recognize only two classes: free or occupied parking space.

Authors in [11] implement a parking detection system based on real-time image processing from video cameras. It divides the system into three sections: image acquisition module, image pre-processing module and image detection module. In the first one, the image is filtered. The detection module is based on the use of a reference image from an empty parking space, without any interference. From this image, converted to grayscale, a comparison is done with successive images. The algorithm obtains the edges of the reference image and compares them with the last captured. Finally, a function is applied to decide whether the reference image and the compared one have similar characteristics. The accuracy of the system is fair (81%).

The solution proposed in [12] uses individual images from a single surveillance camera previously installed in indoor parking lots. This work is also based on reference images from the empty parking lot. The process is slightly different from [11]. The model is generated using Principal Component Analysis, supporting illumination invariance. In addition, textures are also used to detect objects and isolate them from the background. This makes the system much more robust under extreme illumination changes (large trees, street lamps or intense shadows). In areas with high visibility, the number of occlusions between vehicles is minimized and, hence, systems tend to be more accurate. This work has 90% of accuracy, although it can decrease if occlusions are present in the image.

In [13], the classification is done by a 3-layer Bayesian hierarchical detection Framework (BHDF). To avoid occlusion problems with other cars and surrounding objects, this paper uses the full image for processing, rather than determining the state of the parking spaces one by one. To analyze the whole parking space, the scene is divided into 3D cubes. Each cube corresponds to a row of car lots. Once the image is obtained and the row of parking spaces to be processed is defined, the BHDF framework is applied. This framework is composed of three pre-trained models, one for the local classification (observation layer), other with the adjacency model (labeling layer) and, the last, with the semantic layer. The local classification model (input model) can be pixel-based or texture-based.

The work presented in [14] deals with the problem of the sunlight, the dark shadows during the day, and the low light intensity at nighttime. To do this, ParkLotD uses a classifier based on the fuzzy c-means clustering algorithm (FCM) and a hyper-parametric fit with particle swarm optimization (PSO). The algorithm has been tested during different hours of the day and under different weather conditions. However, nighttime use with poor lighting conditions is not advised. At the beginning of the process, it is necessary to define the limits of each parking space, as well as the total number of lots. In the next stage, PCA is applied from the feature vectors obtained from each image. Thanks to a user interface, the operator can correct the algorithm operation and show if any error occurs. Furthermore, the operator can retrain the system when needed. The evaluation process was carried out with 2000 images captured by a camera on a roof. Each image covered 27 parking spaces.

In [15], parking states are determined by the combination of an adaptive background subtraction algorithm for moving object detection (thus overcoming problems of light changes and shadow effects), with speeded-up robust features (SURF) algorithm (robust to scale and rotation changes). The authors propose a solution based on image appearance and the comparison with a reference image where the lots appear vacant. The use of a homogenous transformation to change the point of view facilitates the extraction of the parking model eliminating perspective distortion. For this process, the user must previously select at least four points that define the parking area. Once the transformation is completed, the parking lots are defined. For adaptive background subtraction, the Mixture Gaussian Model algorithm is used. This requires a learning process for updating the distributions obtained by the model through a set of images. Once the background is extracted, the characteristics of the images to be classified are obtained. For this process, the algorithms of SURF feature extraction (invariant to scale changes) and Histogram of Oriented Gradients (HOG) are proposed. The algorithms SVM or KNN can be combined during the classification stage. The authors, through their evaluation with the VI-RAT video database, consider that the best combination of proposed solutions is SURF + SVM, obtaining an average precision of 93%. However, the main problem of this method is the lack of robustness to partial occlusions.

In [16], the aim is not focused on parking area recognition, but on detecting cars and distances by means of vehicles equipped with a stereo vision system. The authors identify three methods to perform this detection: using radar, using active sensors (laser, radar, lidar) and monocular vision, and finally, using only vision. 2D-based systems may not be enough accurate in identification of the vehicles. To solve this drawback, this system processes the 3D image identifying certain characteristics potentially belonging to a vehicle (vertical edges). The algorithm is divided into two main stages: the extraction of three-dimensional characteristics and the detection of the vehicle from them. Extraction of vertical characteristics allows isolating more precisely the obstacles (and vehicles) of the own highway. Two cameras are used to generate 3-D sparse maps. As aforementioned, although this kind of algorithm is not specific for parking lot recognition, they can be applied in this context thanks to the capacity for recognizing vehicles.

Finally, Table 2 shows a schematic representation of some of the most relevant works in parking slot recognition. The table shows various parameters representative of the different solutions proposed. These parameters include hardware required, algorithm error rate, robustness to lighting changes (low ○, medium ◐, high ●), automation level of the recognition and learning process (manual identification of parking lots for training process ○, semiautomatic system allowing manual changes in the training process ◐, fully-automated system ●), robustness to perspective changes (low ○, medium ◐, high ●), robustness to partial occlusions (low ○, medium ◐, high ●), preferred environment for the system use (indoor, outdoor, both), deployment costs (low ○, medium if cameras are already installed ◐, complex infrastructure could be required [satellites, drones, ultrasonic sensors] ●), computational requirements (low [could be high during the training stage] ●, medium◐, high ○), scalability (low ○, medium◐, high ●) and, finally, the way in which the car parks are processed (individually using masks or collectively over the whole area).

Table 2. Vision-based parking lot detection algorithms.

Author	Year	Error rate	Hardware required	Lighting	Learning	Perspective	Occlusions	Coverage	Economic	Computational	Scalability	Grouping
G. Amato et al. [10]	2017	3.0%	Smart-camera	●	⊘	●	●	Both	●	⊘	●	Masks
G. Amato et al. [17]	2016	9.0%	Smart-camera + RPi	●	⊘	●	●	Both	●	⊘	●	Masks
I. Masmoudi et al. [15]	2016	7.0%	Videocamera	⊘	⊘	⊘	○	Outdoor	⊘	⊘	⊘	Masks
J. Suhr et al. [18]	2014	1.9%	AVM and ultrasound	●	●	●	●	Both	○	●	○	Vehicle
J. Jermsurawong et al. [19]	2012	1.0%	Videocamera	●	⊘	○	●	Outdoor	⊘	⊘	●	Full area
H. Ichihashi et al. [14]	2009	7.0%	Videocamera + server	⊘	⊘	○	○	Both	⊘	●	⊘	Masks
Z. Bin et al. [11]	2009	19.0%	Video camera + CPU	○	○	○	○	Both	⊘	●	⊘	Masks
C. Huang et al. [13]	2008	6.0%	IP camera + CPU	⊘	⊘	○	●	Indoor	⊘	⊘	⊘	Rows
G. Toulminet et al. [16]	2006	N/A	Stereo-vision system + CPU	●	●	●	○	Outdoor	⊘	⊘	○	Vehicle
Q. Zhang et al. [20]	2006	20.0%	Satellite images	●	●	●	⊘	Outdoor	○	●	○	Full area
S. Funck et al. [12]	2004	10.0%	CCTV-cameras	⊘	⊘	○	○	Indoor	⊘	●	⊘	Masks
K. Yamada et al. [21]	2001	0.6%	Videocamera	⊘	⊘	●	○	Outdoor	●	●	⊘	Masks
X. Wang et al. [22]	1998	25.0%	Aerial images	●	⊘	●	●	Outdoor	○	●	○	Full area

2.3 Social Crowdsensing

Using software applications to help drivers park their vehicles is not new, even with solutions that include the location, reservation, access and payment tasks all together as in [23]. Nevertheless, the growing popularity of smartphones in the last years, full of sensors and capable of registering geo-positional information in short periods of time, has created new ways of searching for a parking spot giving rise to multiple mobile and web applications. Table 3 shows a comparison between some of the apps for on-street or off-street parking that we can find on the Internet. The selection criteria used to filter the huge catalog that exists nowadays is based on popularity, but also on the special features offered by some of them. Most of the apps are centered uniquely on off-street parking, maybe collecting data from installed sensors or cams. Parkopedia (available worldwide) or wesmartPark (available in Madrid and Barcelona), are different examples of these off-street parking solutions. The latter offers the installation of sensor technology and a management module for owners to easily monitor their business while they obtain data for their app users. While the basic operations in these apps remains the same, we find interesting proposals in the literature that could change the scene someday, i.e. bringing automatic price negotiation capabilities [24]. There are also apps that give information of on-street parking, recommending the best zones to search for unoccupied parking spots, or giving the opportunity to announce when a user is about to move a vehicle, leaving a free space, and in some cases, allowing reservations for these new free spots. Each of these functionalities can use data obtained automatically by cell phones (crowdsensing) or introduced manually by the app users (crowdsourcing). This approach, automatic or manual, is particularly useful for on-street parking, where the scalability of other approaches is more difficult due to the size of the areas to be covered. While off-street parking information can primarily help users reduce the price to pay or select the closest space to their destination, on-street parking information usually has as its main goal the mitigation of the "multiple-car-chasing-single-space" phenomenon [25] derived from the common "blind search" strategy.

Table 3. Smart parking apps

URL	Coverage	Mobile	On-street			Off-street			
			Stats	Pay	Reserve	Location	Prices	Pay	Reserve
www.parkme.com	4200 cities, worldwide	✓	✓			✓	✓	✓	✓
en.parkopedia.com	6308 cities, worldwide	✓				✓	✓	✓	✓
parknav.com	70 cities, USA & Germany	✓	✓						
www.waze.com	N/A	✓				✓			
maps.google.com	25 cities, USA	✓	✓			✓			
www.wazypark.com	Big cities, Spain	✓			✓	✓			
www.aparcandgo.com	Airports/stations 2 cities, Spain					✓	✓	✓	✓
www.parkapp.com	16 cities, Spain	✓				✓	✓	✓	
www.wesmartpark.com	2 cities, Spain	✓				✓	✓	✓	✓
parclick.es	170 cities, Spain & EU					✓	✓	✓	✓
www.telpark.com	+60 cities, Spain & Portugal	✓		✓		✓	✓	✓	
www.e-park.es	13 cities, Spain	✓		✓					
www.parkwhiz.com	+200 cities, USA	✓				✓	✓	✓	✓
www.streetline.com	N/A	✓	✓			✓	✓	✓	✓
www.bestparking.com	+100 cities & airports USA & Canada	✓				✓	✓	✓	✓
www.spotoops.com	Worldwide	✓			✓				
www.justpark.com	+1000 cities, UK	✓				✓	✓	✓	✓
parktag.mobi	7 cities, EU	✓	✓		*a				
www.peertopark.com	Nationwide, Spain					✓	✓	✓	✓
www.aparcalia.com	3 cities, Spain + airports & stations					✓	✓	✓	✓
www.carpling.com	Nationwide, 13 countries				✓				
www.appyparking.com	11 cities, UK	✓	✓	✓		✓	✓	✓	

[a]Planned

When the information is retrieved in an automatic way, it is usually visualized as a statistical estimation, showing the street maps with different colors indicating whether it is easy, approachable or difficult to find parking spots. This coarse grain classification tags can help drivers heading towards zones where parking is more likely to be successful although it obviously does not guarantee an available parking space. This analysis is based on the processing of data (i.e. cell phone signals) in a massive way, generating mathematical models that represent global data, just like it is done to show the traffic density during different hours of the day. ParkMe (owned by INRIX), Parknav, Streetline and Google Maps, are examples of parking solutions that employ machine learning algorithms to process automatically retrieved parking data. Streetline is heavily focused on data analytics and offers different solutions, "Parker" being the one that shows real time parking availability to the user. In the case of Google Maps, it is a new functionality[3] only available in some US cities.

In relation to this big data perspective, we can find research on related tasks that can individually detect parking activity. In order to detect when a car has been parked or when its location is going to become a free spot in a near future, algorithms can focus on activity transition detection, from driving to walking and vice versa, or on the prediction

---

[3] https://blog.google/products/maps/know-you-go-parking-difficulty-google-maps/.

of user destinations, to find out for example if a user is walking towards the vehicle. In [26] we find an example of automatic detection of parking availability through what the authors call "pocketsourcing". Monitoring the activities using phone sensors result in a 94% of correctly predicted parking availability according to their experiments. ParkTAG uses this kind of technology with a so-called "Auto-detection algorithm" and is actually working in seven European cities. The main obstacle is to convince users to keep the app running on background on their devices. At the beginning, there is a lack of users executing the algorithm, which results in incomplete information and low accuracy of the outcome. This is referred to as the *cold start* problem that affects all the crowdsourcing initiatives, forcing us to achieve a critical mass of users for the project to be viable. Other research papers, like [27], focus on the optimization of the overall parking process, considering a cost function that depends on the proximity to destination and the parking price.

Another approach is to focus on the information systems that access the in-car sensor-network present in modern vehicles. Some of the mentioned methods would also apply in this scenario, although with some limitations, as we have access to plenty of data from the car, but not from the user. Therefore, it makes it easier to analyze the movement profile of the car but does not include the user movements, leaving out valuable information for example to predict parking status changes based on user habits or real time monitoring. From a general point of view, the main problem of this approach is the lack of a unified database to share the information coming from vehicles of different brands, which makes it necessary to work with external data from other sources. An example is the BMW on-street parking information system, incorporated in cars of this brand manufactured in Germany or the United States since last year, which uses the INRIX parking information database. BMW is also present in other solutions like Just-Park after a remarkable investment. It is also possible to extend the technology already provided by the car manufacturer installing extra equipment, as the appyparking team is suggesting for example. Installing a M2M dongle in any car will enable the "one click parking" they have developed, allowing semi-automatic payment by just confirming the location on our phone.

Finally, among the apps currently available in Android and iOS marketplaces, we find collaborative apps that allow "manual" sharing of parking information. People announce when they are about to leave, so that other users can drive towards their location and occupy that space. Some of these apps also offer a reservation feature, which forces the outgoing car to publish the information with enough time for its counterpart to arrive, or to wait until the operation is done. The mentioned Parknav app has the "I'm leaving" feature, which is used in their real time parking information system, and Park-TAG claims they will incorporate this feature soon. Others are directly focused on this feature, like spotoops, which is a clear example of the major drawback of this approach, the cold start problem once again. The app is available worldwide, although it does not have enough users, which makes it useless in most cases having no spots offered in the zone we are interested in. Even when the app has a large number of users, it is necessary to implement ways to stimulate the participation of those users and also to encourage them to contribute high quality data [28]. Another app that is trying to overcome this obstacle is Wazypark, first narrowing the scope (it is only available in Spain) and also offering other features to attract more users (like a comparator of gas stations), which seems to be working as the number of users keep growing. Carpling is another collaborative application where parking spots is just one of the assets (car, taxi, train,...) people can share, becoming a collaborative social network. The public condition of on-street parking makes some people disagree with the offer-demand concept of these apps, especially when the user that offers a spot sets a price to pay for it. In 2014, people started publishing negative opinions about this type of technological help that works against those who stay out of these new-born communities, the so-called "freeriders". In fact, the term "jerktech" was created in San Francisco to talk about those startups that make profit by exploiting small businesses and public infrastructure favoring the rich that can pay for their services (MonkeyParking was one of the main victims of this opinion wave). Keeping the monetary aspect aside, there have been some studies showing that parking apps can benefit not only their users, but the entire community, including "freeriders" [29].

3   Discussion

In this paper, the main approaches to the problem of car park recognition have been presented in a general way. Three types of systems have been differentiated: those based on infrastructure, those based on vision and those based on crowdsensing. Each one of them presents advantages and disadvantages that will have to be faced during the next years. In the case of systems based on infrastructure, the main limitation is the cost of implementing such solutions. In addition, the cost of maintenance is also one of the main reasons why many infrastructure-based systems have failed to succeed.

On the other hand, vision-based systems have proliferated in the last decade. Nevertheless, the processing of images still do not obtains good result under certain conditions. The main problems faced by these techniques are illumination changes, partial occlusions and generalization capacity. However, the cost of deployment and maintenance is much lower than in infrastructure-based solutions. The low cost and decreasing size of the sensor itself, the infrastructure and the available software techniques and tools make vision a most promising approach.

Finally, due to the expansion of social systems, crowdsensing techniques are a clear alternative for the future to solve the parking problem. However, because of their lack of maturity and the problems inherent in social systems, their real-life implementation is not yet a reality in the vast majority of cases. The problems of cold start or community involvement require solutions to turn crowdsensing-based systems into an effective alternative or complement to infrastructure or vision based systems.

Acknowledgements. This research is partially supported by the Spanish Economy Ministry and FEDER R&D through the "HERMES–Smart Citizen" project (TIN2013-46801-C4-1-R).

References

1. Bongaarts, J.: United nations department of economic and social affairs, population division world mortality report 2005. Popul. Dev. Rev. **32**(3), 594–596 (2006)
2. Shoup, D.C.: Cruising for parking. Transp. Policy **13**(6), 479–486 (2006)

3. Gallivan, S.: IBM global parking survey: drivers share worldwide parking woes technical report. Technical report, IBM (2011)
4. Mathur, S., Jin, T., Kasturirangan, N., Chandrasekaran, J., Xue, W., Gruteser, M., Trappe, W.: Parknet: drive-by sensing of road-side parking statistics. In: Proceedings of the 8th International Conference on Mobile Systems, Applications, and Services, pp. 123–136. ACM (2010)
5. Bock, F., Eggert, D., Sester, M.: On-street parking statistics using lidar mobile mapping. In: 2015 IEEE 18th International Conference on Intelligent Transportation Systems (ITSC), pp. 2812–2818. IEEE (2015)
6. Lin, T.S.: Smart parking: network, infrastructure and urban service. Ph.D. thesis, Lyon, INSA (2015)
7. Evenepoel, S., Van Ooteghem, J., Verbrugge, S., Colle, D., Pickavet, M.: On-street smart parking networks at a fraction of their cost: performance analysis of a sampling approach. Trans. Emerg. Telecommun. Technol. 25(1), 136–149 (2014)
8. Sanchez, L., Muñoz, L., Galache, J.A., Sotres, P., Santana, J.R., Gutierrez, V., Ramdhany, R., Gluhak, A., Krco, S., Theodoridis, E., et al.: Smartsantander: IoT experimentation over a smart city testbed. Comput. Netw. 61, 217–238 (2014)
9. Ghent, P.: Optimizing performance objectives for projects of congestion pricing for parking. Transp. Res. Rec. J. Transp. Res. Board 2530, 101–105 (2015)
10. Amato, G., Carrara, F., Falchi, F., Gennaro, C., Meghini, C., Vairo, C.: Deep learning for decentralized parking IoT occupancy detection. Expert Syst. Appl. 72, 327–334 (2017)
11. Bin, Z., Dalin, J., Fang, W., Tingting, W.: A design of parking space detector based on video image. In: Proceedings of 9th International Conference on Electronic Measurement and Instruments (ICEMI 2009), pp. 2253–2256 (2009)
12. Funck, S., Mohler, N., Oertel, W.: Determining car-park occupancy from single images. In: 2004 IEEE Intelligent Vehicles Symposium, pp. 325–328. IEEE (2004)
13. Huang, C.C., Wang, S.J., Chang, Y.J., Chen, T.: A bayesian hierarchical detection framework for parking space detection. In: IEEE International Conference on Acoustics, Speech and Signal Processing (ICASSP 2008), vol. 1, pp. 2097–2100 (2008)
14. Ichihashi, H., Notsu, A., Honda, K., Katada, T., Fujiyoshi, M.: Vacant parking space detector for outdoor parking lot by using surveillance camera and FCM classifier. In: IEEE International Conference on Fuzzy Systems (FUZZ-IEEE 2009), pp. 127–134. IEEE (2009)
15. Masmoudi, I., Wali, A., Jamoussi, A., Alimi, A.M.: Vision based system for vacant parking lot detection: Vpld. In: IEEE International Conference on Computer Vision Theory and Applications (VISAPP), vol. 2, pp. 1–8, January 2014
16. Toulminet, G., Bertozzi, M., Mousset, S., Bensrhair, A., Broggi, A.: Vehicle detection by means of stereo vision-based obstacles features extraction and monocular pattern analysis. IEEE Trans. Image Process. 15(8), 2364–2375 (2006)
17. Amato, G., Carrara, F., Falchi, F., Gennaro, C., Vairo, C., Moruzzi, G.: Car parking occupancy detection using smart camera networks and deep learning. In: IEEE Symposium on IEEE Computers and Communication (ISCC), (DI) (2016)
18. Suhr, J.K., Jung, H.G.: Sensor fusion-based vacant parking slot detection and tracking. IEEE Trans. Intell. Transp. Syst. 15(1), 21–36 (2014)
19. Jermsurawong, J., Ahsan, M.U., Haidar, A., Dong, H., Mavridis, N.: Car parking vacancy detection and its application in 24-hour statistical analysis. In: 2012 10th International Conference on Frontiers of Information Technology (FIT), pp. 84–90. IEEE (2012)
20. Zhang, Q., Couloigner, I.: Benefit of the angular texture signature for the separation of parking lots and roads on high resolution multi-spectral imagery. Pattern Recognit. Lett. 27(9), 937–946 (2006)
21. Yamada, K., Mizuno, M.: A vehicle parking detection method using image segmentation. Electron. Commun. Japan 84(10), 25–34 (2001). (Part III: Fundamental Electronic Science)

22. Wang, X., Hanson, A.R.: Parking loT analysis and visualization from aerial images. In: Fourth IEEE Workshop on Applications of Computer Vision (1998)
23. Hodel, T.B., Cong, S.: Parking space optimization services, a uniformed web application architecture. In: ITS World Congress Proceedings, pp. 16–20 (2003)
24. Chou, S.Y., Lin, S.W., Li, C.C.: Dynamic parking negotiation and guidance using an agent-based platform. Expert Syst. Appl. **35**(3), 805–817 (2008)
25. Wang, P., Hunter, T., Bayen, A.M., Schechtner, K., González, M.C.: Understanding road usage patterns in urban areas. arXiv preprint (2012). arXiv:1212.5327
26. Nandugudi, A., Ki, T., Nuessle, C., Challen, G.: Pocketparker: pocket sourcing parking loT availability. In: Proceedings of the 2014 ACM International Joint Conference on Pervasive and Ubiquitous Computing, pp. 963–973. ACM (2014)
27. Geng, Y., Cassandras, C.G.: New "smart parking" system based on resource allocation and reservations. IEEE Trans. Intell. Transp. Syst. **14**(3), 1129–1139 (2013)
28. Hoh, B., Yan, T., Ganesan, D., Tracton, K., Iwuchukwu, T., Lee, J.S.: Trucentive: a game-theoretic incentive platform for trustworthy mobile crowdsourcing parking services. In: 2012 15th International IEEE Conference on Intelligent Transportation Systems (ITSC), pp. 160–166. IEEE (2012)
29. Chen, X., Santos-Neto, E., Ripeanu, M.: Crowdsourcing for on-street smart parking. In: Proceedings of the Second ACM International Symposium on Design and Analysis of Intelligent Vehicular Networks and Applications, pp. 1–8. ACM (2012)

Impact of Protests in the Number of Smart Devices in Streets: A New Approach to Analyze Protesters Behavior

Antonio Fernández-Ares[✉], Maria Garcia-Arenas, Pedro A. Castillo,
and Juan J. Merelo

ETSIIT-CITIC, University of Granada, Granada, Spain
antares@ugr.es

Abstract. Measuring protests is an area of interest, as the amount of protesters is proportional to the success of the protest. Nevertheless, current methods of measurement are in counting heads in photos. In this paper using smart devices detection in a protest to measure the amount of people in the journey it is proposed. In order to do so, the *Mobywit* System is used, having been employed with success in the monitorization of vehicles and persons in Smart Cities scope. This system tracks the smart devices using their WiFi communications. Gathered data measures the number smart devices taking part in the protest, and so the number of participants.

Keywords: Smart cities · People monitoring · People tracking · Wifi monitoring · Protests

1 Introduction

A protest, protest march or rail implies a huge number of citizens in the streets proclaiming a complaint to the government or their support for a specific cause. Usually the success of a protest is measured using the number of people that flow along its route. The methods employed for these measures are rudimentary, or even in many cases, of low accuracy. That implies that for a protest the number of protesters might be different for the government, for the organizers, for the media and for anyone who wants to estimate.

Currently, the most precise way of measuring a protest implied analyzing aerial photographs and counting heads. In a protest with a huge number of people, this is a titanic task that usually is made by humans. That is why an common practice is to split the photo in squares with the same area, and counting the heads only in one of the squares. Then, the number of protesters is got just multiplying. This method is called Jacobs's Method [11].

This method presents several limitations: the concentration of people is not the same in all the squares, it is hard get squares of the same area using just

ⓒ Springer International Publishing AG 2017
E. Alba et al. (Eds.): Smart-CT 2017, LNCS 10268, pp. 75–85, 2017.
DOI: 10.1007/978-3-319-59513-9_8

photos because photos have perspective and projection areas and because the urban furniture is not covered by the problem. And the most important problem, implies humans in the process of counting heads. There have been attempts to automatize the process [1], but they have not been successful.

Other approaches have focused in social networks, using the sent tweets from a particular place [2] as measurement of the success of the protest. Or even is stating to talk about online protest, where is more important the impact of the protest in the social networks than on the streets [6].

Estimating the number of persons in a crowd is hard and have many interest [15], not limited to protests environment. However, today's systems are not able to measure crowds, such as those present in protests. In this sense, infrared beams [8] are widely used to counting people walking in a limited bounded places, for example a gate or a wheel, but they can not be applied to a open place like a street. Thermal counters [7] and video cameras [4,14] present problems in the detection of people if the image is not clear, that means, present obstacles and crowded people, something common in protests.

Thus, this paper presents a new method that can be useful to study and measure attendance level to a protest. Proposed method takes advantage of the smart devices that the protesters wear in the protest, and allow to count, monitor and track the movement of the smart devices, and therefore of the protesters that carry them. Thus, the communications emitted by the devices are capture and processed by the proposed a system, called *Mobywit* [3].

The rest of the work is structured as follows: next Sect. 2 briefly presents the *Mobywit* system. Then, the problem studied is introduced in Sect. 3, where detailed information about protests and about the *Mobywit* nodes used in this work is presented. Section 4 shows the obtained results of the experiments carried out. Finally, Sect. 5 presents the reached conclusions, followed by future works in this research line.

2 Tracking People Using WiFi Signals

The idea of tracking people using their smart devices has been exploited because more and more people use them every day. It has also become a very profiting research area [10,12,13]. Xi [16] and Fierro et al. [5] demonstrated how tracking people with their personal WiFi devices can be used to count people inside a building without requiring additional infrastructure and with great accuracy.

In this work, a number of *Mobywit* nodes are placed in the streets, more specifically within the traffic lights as Fig. 1 shows. These nodes are provided with WiFi antennas that are able to operate in monitor mode. In this mode, all the traffic received from the wireless network can be monitored without having to associate to an access point or ad hoc network, unlike promiscuous mode.

Each device emitting frames, that are received by the antennas of the node, is uniquely identified by the physical address or *MAC Address* that is used to network interfaces for most of IEEE 802 network technologies. *MAC Addresses* are assigned by the manufacturer of the network interface controller (NIC) and are stored in its hardware, such as the NIC's read-only memory or some other firmware mechanisms. That implies that the MAC Address of the smart devices can not be changed, so a particular device shows always the same address.

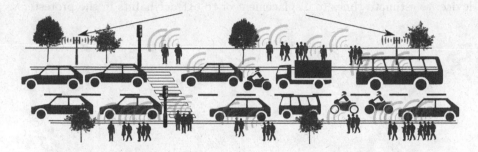

Fig. 1. Wireless tracking of people, using the signals that their smart devices are emitting in their communications.

The smart devices are always looking for open or known WiFi networks, that implies send frames beacons. These frames are caught by the *Mobywit* nodes, taking note of the *Mac Address* of the device that send the frame. Inasmuch as the smart devices are periodically sending these frames, the *Mobywit* nodes can approximate how much time elapsed each device near of the node, using the time of the first and the last frame emitted by the device that have been catched up by the *Mobywit* node.

This process it is totally legal, because the protocols Bluetooth and WiFi allow to search for near devices, being one of the main service of the protocol. As the *Mac Address* of a device it is not linked (unlike a number plate or phone number) it is impossible to know the owner or carrier of the device tracked.

A scenario with several *Mobywit* nodes working together, allows to measure the time needed to go from a place to another or track the displacement of the people. In a protest, this system offers information about the number of protesters in the streets, the time that they spend outside and the speed of the crowd.

3   Analyzing Smart Devices in Protest

The collaboration of the Mobility Area of the Local Council of Granada City allowed us to install several *Mobywit* nodes in the city center. These nodes, are located near the starting point of the protest, as Fig. 2 shows[1].

Three protests and a concentration have been carried up in the area near to the *Mobywit* nodes. Protests and the concentration day were Sundays from 12:00 onwards. These four days have been compared with four regular Sundays during the same time period.

---

[1] Generated using the tool described in [9].

The existence of an increment of the number of devices, and therefore of people, in the moments of protests, shows that the *Mobywit* System can be useful in the analyses of protests. These analyses can be focused in the counting of protesters or, thanks to the process of the system to identify unequivocally each device, to estimate times to displacement or to extract habits in the protesters.

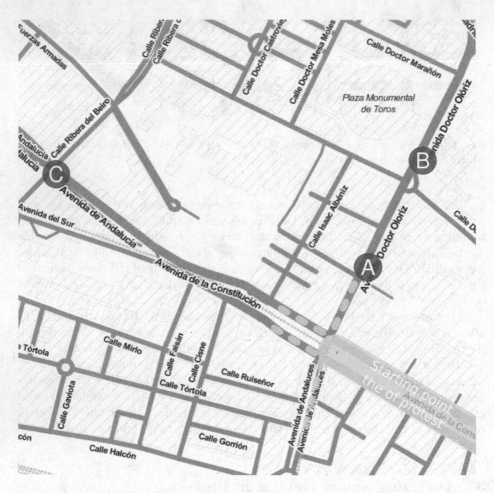

Fig. 2. Map showing the location of each *Mobywit* node named *A*, *B* and *C*. In red, the tail (or starting point) of the protest (Color figure online).

4   Experiments and Results

This section presents the experiments carried out using the data collected on the manifestation days and on the regular days.

4.1   Number of Devices/People by Hour

The *Mobywit* System can be used to count for the number of devices (people) that have passed through a particular site, close to the node, in a time period. Figure 3 shows the number of devices tracked by hour for each day.

On regular days, similar number of devices have been detected in the three nodes, with no significative difference according to the ANOVA test, as Fig. 4 shows. Before doing these tests, it has been proven that the number of people by hour conforms to a normal distribution. In point A, the closer to the protest, there are significative differences between the regular days and the protest days P2 and P3. In point B, there are strong differences between the regular days, and the protest and concentration days. In point C, there are not relevant differences because it is the farthest point of the protest.

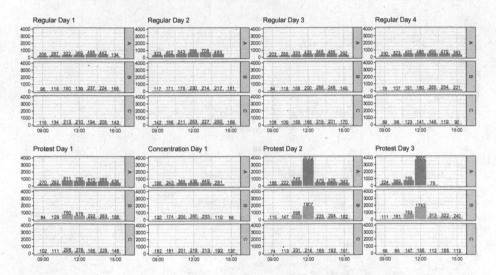

Fig. 3. Number of devices tracked by hour, in the 8 days studied, in the 3 node locations.

That means that a protest in the city center has a clear impact in the number of smart devices in the street, because there are more people in the areas near to the nodes.

Increments in amount of people in streets

As the system it is not exhaustive, the precise way to present information about the number of devices detected is to show the increment of devices compared to the regular days allowing to check the important hours of the protest, and have an approximate measure of the amount of people additional compared to a regular day.

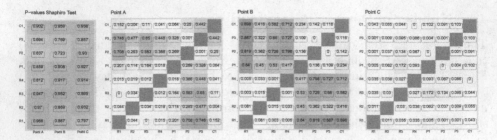

Fig. 4. P-values of the Shaphiro test of normality. Thereafter for each point, p-values of the ANOVA test of each day: regular, protest and concentration.

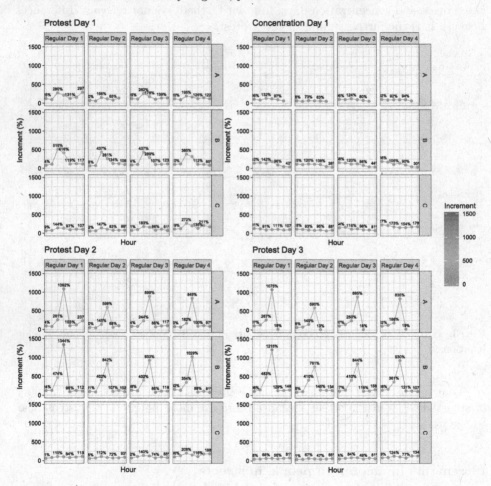

Fig. 5. Increments of the number of devices detected by hour based on the regular days of each protests day.

Figure 5 shows the increment of detected devices in the protest days. In the protest days 2 and 3, there is an increase around 10 times more devices in the hours near to the protest. Paying attention to the amount of people, it can be extrapolated that have passed, in turn, 10 times more people than usual. In the other hours, there are not a huge variation in the number of devices detected.

Taking into account the protest day 1, the increment is lower than the other protest days, around four or five times more devices. In the concentration day, there are not an increment significative.

This information can be useful to measure the impact of the protest in the city. As it can be quantified the amount of extra people due to the manifestation, it can be derived that at that time, on regular days, that people wouldn't have been in that locations.

4.2   Minute by Minute Analysis of the Protest

Cause the system identify each device with a period of time, it is possible to rebuild the protest, knowing in each moment how many devices (people) there are in the area near to the node. Figure 6 shows the number of devices tracked in

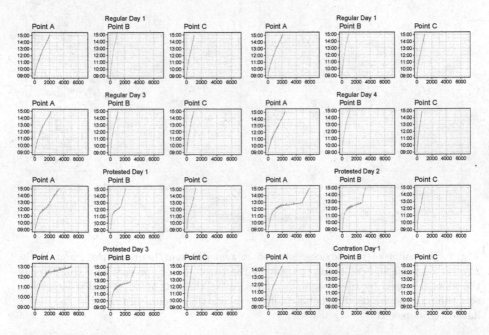

Fig. 6. Number of devices detected by hour in each point. In the y-axis, the time of the day from 9 am to 15 pm. Each point in the x-axis represent a device tracked, where the x-value represent that the device was the number x in be detected. And that device was detected between the time shown by the two y-values. That means, the width of the line of points represent the amount of time that the each device have been in the point.

each second by the system during the protest time. In each subfigure a line for each device is shown (the height is proportional to the time that the device are detected in that node). Using that information, it is easy to infer the approximate number of devices that in a particular moment was in the place.

Hence, the system is able to know how much and what devices were being detected at any given time getting the state of the detected devices and allowing reproduce, in some sense, the protest situation. Figure 7 shows the devices detected in the *Point B* the *Protest Day 2* at two different hours. At 10:00 the protest had not started, so the number of devices in the area are low. But at 12:00, star time of the protest, the number of devices in the area have increased until ten times more devices.

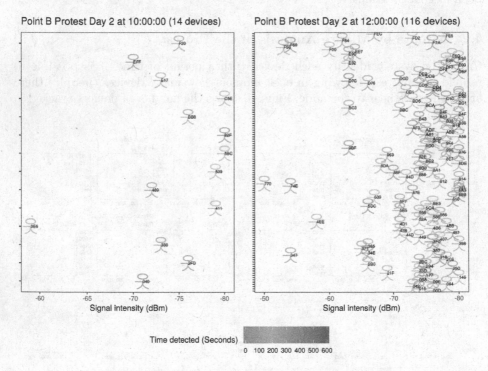

Fig. 7. Snapshot of the devices detected in two different hours in the Point B the *Protest Day 2*. Each device is represent by the three more representative character of the identifier used by the system, calculated from the MAC ADDRESS and it is used to sort them in the y-axis. The point color represents the amount of time that the device have been detected.

This information is useful as if we observe in real time while the protest flows as if we want to study the data later. In addition, it is capable of generating a step-by-step animation that shows the concentration of people along the time.

4.3   Recurrence Analysis

Because the system identify each device, it is possible to know if a particular device is habitual in some scenario or its occurrence is exceptional. For example, Fig. 8 shows the common devices detected in the four regular days. These devices has been detected several days in the same interval, so their occurrence it is habitual in all the other days.

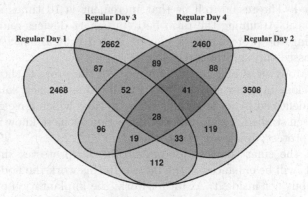

Fig. 8. Number of devices tracked in regulars days.

If those devices are removed, it can be studied the devices that have attended all protests as shown in Fig. 9. The amount of devices that have been detected in protests of day 2 and 3 are relevant. Around a 15% of the devices detected in the protest day 3 was detected too in the second protest.

This information can be useful to understand if the protesters are assiduous.

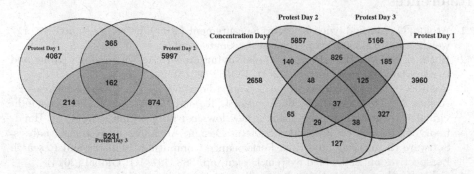

Fig. 9. Number of devices tracked in protest days, including or not the concentration day. This figure shows the number of devices that have been detected in the same day, that means they have attended the protest, excluding the regular devices.

5   Conclusions

In this paper, the impact of protests in the amount of smart devices in streets has been studied. It has been observed a statistically significant increase of the number of these devices during the protest, comparing them with regular selected days. This fact is an evidence of the viability of the method for measuring protests including along the way several Mobywit monitors.

Thanks to this information, the increment of walking people along the streets can be analyzed. The results tell us that increment is 10 times more devices than regular days. Assuming the ratio between smart devices and people it is invariable during the protest occurs, can be stated that about 10 times more people have passed through the streets.

Moreover, Mobywit is able to get snapshots of the protest, showing devices which passed close in a certain moment. Finally, due to each smart device is unambiguous identified by its MAC, the smart devices which have gone to every protest have been studied. The results in this section let us to know if the devices whose owners protest always are the same.

The use of the smart devices for measuring the protesters in an area of special interest, will be enhanced in the future. In this work the nodes were close to the protest but not inside it. As future work, the implantation of a detection Mobywit node in the route of the protest is proposed.

Acknowledgements. This work has been supported in part by the Ministerio español de Economía y Competitividad under project TIN2014-56494-C4-3-P (UGR-EPHEMECH) and PRY142/14 (que ha sido financiado íntegramente por la Fundación Pública Andaluza Centro de Estudios Andaluces en la IX Convocatoria de Proyectos de Investigación) (The description in Spanish is mandatory.). We also thank the DGT and local council of Granada city, and their staff and researchers for their dedication and professionalism.

References

1. Atolon, G.: Method lynce for crowd measurement (2011). https://web.archive.org/web/20110915235904/http://lynce.es/es/metodo.php
2. Botta, F., Moat, H.S., Preis, T.: Quantifying crowd size with mobile phone and twitter data. R. Soc. Open Sci. **2**(5), 150162 (2015)
3. Castillo, P., Fernández-Ares, A., García-Fernández, P., García-Sánchez, P., Arenas, M., Mora, A., Rivas, V., Asensio, J., Romero, G., Merelo, J.: Studying individualized transit indicators using a new low-cost information system. In: Handbook of Research on Embedded Systems Design, Advances in Systems Analysis, Software Engineering, and High Performance Computing. Industry and Research Perspectives on Embedded System Design, pp. 388–407. IGI Global (2014)
4. Celik, H., Hanjalic, A., Hendriks, E.: Towards a robust solution to people counting. In: 2006 IEEE International Conference on Image Processing, pp. 2401–2404 (2006)
5. Fierro, G., Rehmane, O., Krioukov, A., Culler, D.: Zone-level occupancy counting with existing infrastructure. In: Proceedings of the Fourth ACM Workshop on Embedded Sensing Systems for Energy-Efficiency in Buildings (BuildSys 2012), NY, USA, pp. 205–206 (2012). http://doi.acm.org/10.1145/2422531.2422572

6. González-Bailón, S., Borge-Holthoefer, J., Moreno, Y.: Broadcasters and hidden influentials in online protest diffusion. Am. Behav. Sci. **57**(7), 943–965 (2013)
7. Hashimoto, K., Morinaka, K., Yoshiike, N., Kawaguchi, C., Matsueda, S.: People count system using multi-sensing application. In: 1997 International Conference on Solid State Sensors and Actuators, TRANSDUCERS 1997, Chicago, vol. 2, pp. 1291–1294, June 1997
8. Hashimoto, K., Kawaguchi, C., Matsueda, S., Morinaka, K., Yoshiike, N.: People-counting system using multisensing application. Sensors Actuators A Phys. **66**(1–3), 50–55 (1998). http://www.sciencedirect.com/science/article/pii/S092442 4797017159
9. Kahle, D., Wickham, H.: ggmap: spatial visualization with ggplot2. R J. **5**(1), 144–161 (2013). http://journal.r-project.org/archive/2013-1/kahle-wickham.pdf
10. Kostakos, V.: Using bluetooth to capture passenger trips on public transport buses. arXiv preprint (2008). arXiv:0806.0874
11. McPhail, C., McCarthy, J.: Who counts and how: estimating the size of protests. Contexts **3**(3), 12–18 (2004)
12. Morrison, A., Bell, M., Chalmers, M.: Visualisation of spectator activity at stadium events. In: 2009 13th International Conference on Information Visualisation, pp. 219–226. IEEE (2009)
13. Nicolai, T., Kenn, H.: About the relationship between people and discoverable bluetooth devices in urban environments. In: Proceedings of the 4th International Conference on Mobile Technology, Applications, and Systems and the 1st International Symposium on Computer Human Interaction in Mobile Technology, pp. 72–78. ACM (2007)
14. Schofield, A., Mehta, P., Stonham, T.: A system for counting people in video images using neural networks to identify the background scene. Pattern Recognit. **29**(8), 1421–1428 (1996). http://www.sciencedirect.com/science/article/pii/0031 320395001638
15. Stanton, W.J.: Fundamentals of Marketing. McGraw-Hill, New York (1967)
16. Xi, W., Zhao, J., Li, X.Y., Zhao, K., Tang, S., Liu, X., Jiang, Z.: Electronic frog eye: counting crowd using wifi. In: INFOCOM, 2014 Proceedings IEEE, pp. 361–369, April 2014

Logistics Support Approach
for Drone Delivery Fleet

Asma Troudi[1(✉)], Sid-Ali Addouche[1], Sofiene Dellagi[2],
and Abderrahman El Mhamedi[1]

[1] QUARTZ Laboratory EA 7393, IUT de Montreuil- Paris8 University,
140, rue de la nouvelle, 93100 Montreuil, France
{a.troudi,s.addouche,a.elmhamedi}@iut.univ-paris8.fr
[2] LGIPM Laboratory, URF MIM, ILe de Saulcy Metz, 57000 Metz, France
sofiene.dellagi@univ-lorraine.fr

Abstract. This paper treats a drone delivery parcel's problem in an
urban area. Drone delivery has emerged as a potential way in the imme-
diate future to deliver parcels in the urban area, especially in last-mile
delivery. This new configuration of parcel delivery highlights the improvement of the application of Unmanned Aircraft Vehicles in the civil use.
The future challenge underlying this application is not so much the design
of drones for parcel delivery, but the logistics support of a massive fleet
of drones with a mission to deliver at least hundreds of parcels a day in
a dense urban area. We hereby treat this issue by focusing on Logistics
Support System. In this paper, we propose a Post-Production Logistics
Support Analysis to cover the exploitation phase of a drone delivery
operator.

Keywords: Transportation logistics · Logistics support · Intelligent
transportation systems · Drone delivery

1   Introduction

The development of Unmanned Aircraft Vehicles (UAVs) has been improved
over the last two decades. This speed growing has given the ability to UAVs not
only to support dangerous military missions where pilot operations can be risky
but also to be integrated into civil activities.

Civil and public use of UAVs is then becoming more widespread and is touch-
ing many other domains like e-commerce. Some companies are already experi-
menting drones as a mean of delivery. For example, we can mention DHL with
PaketKopter [17], Amazon with Amazon PrimeAir [21], Google with Project
Wing [19] and recently GeoPoste with Geodrone [1]. With this novelty, and
through these applications, we speak about intelligent transportation. The integration of the intelligent transportation reperents nowadays one of the smart
city'elements [15]. That is what the various previous programs try to set up.

These programs present a high potential in emerging markets and rapid
urbanization context. Furthermore, these commercial programs establish a new

© Springer International Publishing AG 2017
E. Alba et al. (Eds.): Smart-CT 2017, LNCS 10268, pp. 86–96, 2017.
DOI: 10.1007/978-3-319-59513-9_9

framework for using UAVs. In addition, this new framework illustrates a part of the smart city definiton according to Washburn et al. [20]. The civilian use of UAVs fleet in last mile delivery introduces new issues to handle.

In this paper, we focus on the Logistics Support for UAVs fleet. We will present Logistics Support approach for UAVs fleet in urban parcels' delivery. In the next section, as a first point, we will present a literature review development of the UAV from the military application to the civil application. We also introduce the Logistics Support approach for the aerial vehicle. In the third section, we propose a post production Logistics Support Analysis for a fleet of UAV in a delivery context.

2 Literature Review

To better understand the issues and difficulties in organizing the missions and Logistics Support of UAVs, it is helpful to understand their characteristics and differences.

Actually, the improvement of UAVs requires many high-level calculation items. UAVs specialists rise the size, the flight duration, altitude, range and the way that these devices take-off and land. We distinguish many types of UAVs: vertical take-off and land UAVs called VTOL, small UAVs, HALE, LALE, MALE, [6,22]. These classifications depend on the flight altitude, the speed, and the endurance. With the expand of the UAVs'use, authorities have established regulations to manage the exploitation of these devices according to many specifications like weight and also technical characteristics as engine's power. Same UAVs require authorities approval for each flight. In fact, the works of Mueller [14] and Maddalon et al. [13] identify many categories like I, II, III according to USA authorities (FAA: Federal Aviation Administration).

In agreement with the French regulation, there are seven categories. These categories depend on the maximum takeoff weight, the operation, the case of model aircraft. For example, the UAVs with which DHL made its first delivery flight belong to the category E (Drone for civil applications with a global weight less than 25 kg) with only scenarios no. 1 and no. 2. The French legislation shows the different flight scenarios and their requirements, [9].

We notice that the majority of these scenarios require only an inhabited flight zone. Nevertheless, the legislative authorities improve the law text continually following the expand civilian activities. As referred to in the introduction, we concentrate on the civil use of UAVs and especially in parcel delivery for the commercial market. This type of delivery has been developed for the last decade. The growth of the transportation activity is estimated about 37.4 billion euros in 2008 according to [8]. Maddalon et al. [13] highlight the potential of using delivery UAVs in urgent cases seems to be attractive especially for one-day and same-hour delivery. We focus on the way that companies have to manage its massive fleet taking into account their missions and how to establish an appropriate Logistics Support (LS).

The new use configuration of UAVs in the delivery application will prompt new constraints to the UAVs operators and demand a special logistics support.

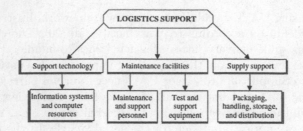

Fig. 1. Elements of Logistics Support for a product according to Murthy et al. 2004

The establishment of the Logistics Support represents a critical phase to provide the park of all the necessary support tools for an efficiency fleet.

Logistics Support system is fundamental to handle and manage a massive fleet of UAVs. Feyman et al. [11] note that the Logistics Support is a balance between life cost, performance, and operational availability.

Logistics Support appeared in military application due to the fact of the complexity of military systems and severe use condition. In the literature, we find various definition of the Logistics Support. Chiesa et al. [3] introduce the Logistics Support system as a concept that helps system to operate properly in a satisfactory way For the french association of system engineering (AFIS), the logistics support is a set of process and resources called elements of logistics support that insure the operational maintenance of a complex system [10].

According to Murthy et al., the elements of Logistics Support is a set of support technology, maintenance facilities an supply support as shoafwed in Fig. 1. The application of the Logistics Support in military activities, plays a significant role. Shukla et al. [18] interested in integrated logistics for fighter aircraft in a Product Life Management environment. This approach assist is defining the performance and the reliability objectives for the maintenance program with the necessary support.

Moreover, the works of Omidshafiei and Agha-Mohammadi [16] focus on the maintenance facilitates and in particular UAV health status in a context of delivery missions. In this works, the authors develop a decision-making tool for each UAV: the vehicle decides in wich base will return after the delivery mission and when to perform a repair action in function of its three health status (high, medium and low). We can conclude that the application of the UAV status influence the UAV operations and missions.

The Logistics Support can also be applied in the industrial warranty domain. In the works of Diaz et al. [7], authors present an outline the role of the logistic support to increase process efficiency in the warranty management. They propose a framework that shows the role of logistics support in the warranty management.

In the context of the maintenance as a part of the logistics support system, Dumas [4] introduces 3 Technical Operation Levels (TOL) of maintenance in function of the task frequency and complexity. Otherwise, there are new para-

Table 1. The supply levels

	TOL1	TOL2	TOL3
Industrial supply level	X	X	X
Operational supply level	X	X	

meters instead of the Technical Operation Levels: the Industrial or Operational Supply Level. These levels indicate the means that will be available to do tasks.

The Industrial Supply Level involves industrial partners and contractors. These partners could cover the 3 levels of the Technical Operation. The Operational Supply Level implies users and includes the first and the second level of the Technical Operation Levels as shown in Table 1. On the other hand, Chiesa et al. [3] show where the maintenance task for UAV in the army should be done: The Depot-Level Maintenance or the Organization Field Level Maintenance. The first level ensures the repair the Line-Replaceable Unit (LRU) and the piece parts. The second level is divided into two sublevels: the shop maintenance which identifies and replaces the LRU faults. The second sublevel is the flight line maintenance which is charged in doing inspections and maintenance routine operations.

Karaaugacc et al. [12] introduce three structure to indicate where different maintenance tasks should take place: Operational Level (O-Level), Intermediate Level (I-Level) and Depot level (D-level).

The O-Level is the daily task and operations including inspections done on the flight line. The I-Level is an off-equipment repair. It is more accurate and complicated operations which should be done in a base shop. The D-Level is specialized in heavy maintenance services as rebuild, overhaul and extensive modifications in the depot.

Referring to previous works about the applications of maintenance for aircraft and military UAVs, we conclude that maintenance could be applied following three parameters: maintenance levels, maintenance stockholder, and maintenance infrastructure. Are these parameters answers about 3 question? How many maintenance levels should we have? Who holds these maintenance operations and where could they do them?

3   Proposal of Generic Approach to Making Use of UAV-Logistics Support

As shown in the previous section, the Logistics Support takes an important place in the management phase of complex systems and especially aeronautic vehicles.

The identification of the fleet status, the health conditions or also maintenance facilities, helps to support the system.

To cover all the Logistics Support aspect in the life cycle, it is mandatory to use a standard to rule the Logistics Support system.

90 A. Troudi et al.

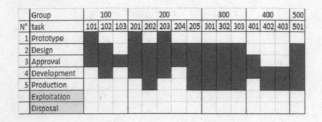

Fig. 2. Logistics Support Analysis (L.S.A)

Works of Bardou [2] detail the use of the Logistics Support Analysis (L.S.A) established through the standard MIL-STD-1388-1A. This analysis was constructed around the development of the system design and its production. This analysis is divided into five groups, [5]:

– 100: Program Planning Control
– 200: Mission and Support Definition
– 300: Preparation and Evaluation of Alternatives
– 400: Determination of Logistic Support Ressource Requirments
– 500: Supportability Assessment

The L.S.A covers initially 5 phases in the life cycle of a system: Prototype, design, approval, development and production or manufacturing, Fig. 2. In fact, we notice that the L.S.A proposed by STD-1388-1A uncovers the exploitation phase and the disposal phase, as shown in Fig. 2. For example, the group 500 (the supportability group) evaluates the ability of the system to be supported during the development phase and the exploitation phase. Besides, the task 501 in this group focus on testing, evaluation of the supportability with a set of strategies plan and assessment of the various elements of support and does not supervise these support elements after the exploitation.

As a consequence, the MIL-STD-1388-1A with its L.S.A does not guide us to control the Logistics Support adapted to exploitation phase applied in our case: fleet of delivery UAVs.

Once the UAV is commercialized, its operating manual describes, in a global way, the strategies of the Logistics Support adapted in function of its design. These instructions do not take into account its exploitation or disposal phases. This assessment leads us to think about a Logistics Support system more adapted to the various requirements and the different constraints during the exploitation phase. In fact, all the L.S.A steps finish in the production phase but what about the exploitation?

The particularity of the UAV activity that we propose, needs a Logistics Support on the shelf. This type of support is in function of the exploitation of the UAV and the type of the activities aside from what was provided during the acquisition phase (operating manual). To resolve this issue, we should establish an approach to drive a Logistics Support System on the shelf.

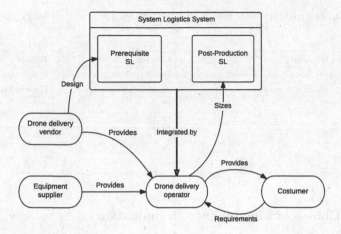

Fig. 3. Configuration of drone delivery park for a drone delivery operator

We remind that we want to manage a fleet of UAVs in an urban delivery context. Our aim is to provide a Logistics Support system which fit with this activity and taking into account all the constraints due to UAV's structure (autonomy, speed), missions (duration, delivery time) and also regulation (range, speed, and altitude).

The identification of the configuration of the UAV delivery operator helps us to identify all the stakeholders related to the operator. Once the contractor provides the operator of UAVs and also a required Logistics Support through the operating manual, the UAVs delivery operator evaluates the exploitation rate of its UAV fleet through the customer demand. This indicator helps the operator to equilibrate the Logistics Support system in function of its activities as shown in Fig. 3.

The L.S.A establishes the Logistics Support related to the design and the manufacturing of the UAV. However, the Post-Production Logistics Support is not covered by this analysis. In the next step, we propose a Post-Production Logistics Support Analysis (P-P.L.S.A) to govern the post-production phases: from the exploitation to the disposal phase.

The objective is to establish a P-P.L.S.A inspired by the L.S.A to keep a coherence between the prerequisite SL and the P-P.L.S. In this context, we will suggest a P-P.L.S.A adapted to a delivery UAVs' fleet, and we focus especially on the exploitation phase as a major phase in the life cycle of the UAVs' fleet. This analysis considers a commercialized drone delivery as a reference and tries to detail what requirement the operator should include. To keep a coherence between the operations manual and our proposal, we will be based on the L.S.A and extrapolate same tasks. We suppose that the L.S.A had already done before the commercialization of the UAV. We search what we could project to the rest of the phases in the life cycle this fleet. After studying the L.S.A, we are interested in two groups: the group 400 Determination of Logistics Support Ressource Requirments and the group 500 the Supportability Assesment.

Table 2. The characteristics of TOLs performed by drone delivery operator

TOL levels	TOL1	TOL2	TOL3
Operation	Replace	Replace	Repair
Means	Without tools or Standard tools	Dedicated and test tools	Industrial plant
Knowledge	Basic knowledge	Specific knowledge	Expert knowledge
Complexity	*	**	***
Accessibility	*	**	***
Security	*	**	***
Procedure	*	**	***
Training	*	**	**

3.1 The Choice of Maintenance Organization

We propose to extend the tasks of group 400 to cover the exploitation phase for the fleet of UAVs. This group identify the Logistics Support exigencies and the impact of eventual modification of the system.

The task 401 (Tasks Analysis): analyze the required maintenance tasks and operation by identifying Logistics Support elements for each task. This analysis should also be applied in the exploitation. It should cover the evolution of the utilization of the UAVs' fleet and the evolution of its health status. In this step operator can plan the personnel qualification in function of the task complexity and frequency.

Through previous works related to the Logistics Support for the military vehicle (aircraft, UAV), we establish the maintenance organization for drone delivery operator. The operator is responsible for maintenance operations only for the Operation Supply Level. In this level of responsibility, the operator assures two Technical Operations Levels of maintenance: TOL1 and TOL2. These activities cover operations related to the LRU as basic test or replacement. They do not require a high qualified personnel or specific tools for maintenance operations.

As an example, the drone operator can do the battery replacement as TOL1. This operation is realized without tools and also a specific knowledge. Table 2 presents the TOLs performed by the drone delivery operator.

To resume all the maintenance aspect related to the drone delivery operator the Fig. 4 represents the different position occupied by the operator (in blue). The graphic shows the leeway area performed by the drone operator to drive its Logistics Support system. This representation leads to identify the essential support resources related to the maintenance organization.

3.2 Identify the Eventual Modifications in the Logistics Support System

During the exploitation phase, the UAV fleet needs same modification due to the development of the activity.

For that, we will explain how we handle these evolutions in our Logistics Support System.

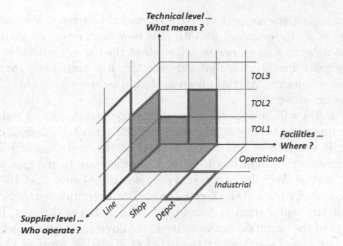

Fig. 4. The implementation of the maintenance for the drone delivery operator (Color figure online)

The task 402 (Early Fielding Analysis): studies the impact of the renovation or the introduction of a new system on existing systems. During the life cycle of the system, same equipment might be modified or modernized. We can take the example of the UAV battery that can be modified to have more autonomy or less weight. This modification impacts the mission plan or the charging time. Besides, the amendment of this component might modify the procedure involved this task. In general, the modification of same component impact related operations process.

The task 403 (Post-Production Support Analysis): analyzes the life cycle support of the system before its production. This task should guide the system during the rest of its life cycle. Then, we will extend it during the Post-Production phases to organize the future modifications and new support system that may be used as a consequence of these modifications. We will keep the same example as cited in the previous tasks. The introduction of a new battery with new technical characteristics may require a new training program for the personnel. The same modification could impact the process. Drone delivery operator should handle the modification and update its Logistics Support system through the P-P.L.S.

3.3 Update the Logistics Support System

In the Post-Production Logistics Support Analysis, we want to emphasize the role of the Supportability Assesment related to Group 500 in the L.S.A.

Through the task 501 (Supportability Test, Evaluation, and Verification) this group controls and tests the achievement of specified requirement for system supportability.

We extend this task for our P-P.L.S.A in order to verify all the modification that might be done during the exploitation phase. The introduction of a new battery may impact the support system that we already have. We should test if the charging post, for example, fit with this new battery or we should modify them. The storage area zone are adapted to stock the battery with new volume or not, battery control software is updated with this new technology, or we should purchase a new software. All this question should be covered in this task during the exploitation phase.

The P-P.L.S.A will be applied in the drone operator base and warehouse. It will also be impacted by the delivery activity the planning mission, charging,.. etc. In fact, it is worth to add a task related to the evolution of the activity. As a consequence, we add in this P-P.L.S.A another task in the group 500. It is the task 502 called Activity Analysis. This task makes sure that the Logistics Support used are adapted to the fleet's needs. The Activities Analysis takes into consideration the exploitation rate and the health status of the fleet. It is a continual control of the evolution of the activities realized by the fleet. Adding other vehicles, for example, may impact the existing Logistics Support. As a consequence, the activities analysis helps the operators to provide their fleet to the adapted Support System. In this task, we answer the following question: does the Logistics Support means are adapted to the system activities? The activity analysis takes into account the activity type, its requirements, and constraints. This information represents an input to size the Post-Production Logistics Support Fig. 3. To summarize, The P-P.L.S.A accounts for a specific continuation of the L.S.A applied for a complex system as a UAVs'fleet in a context of urban delivery.

Drone delivery operators want to manage their fleet adequately. To equilibrate the activity rate of the UAVs fleet and all the support operations, they should provide a necessary number of UAVs in order to satisfy the costumer's demand. In fact, to establish a rigorous Logistics Support related to a particular use of a system in general, we should start by the identification of the activity to determine the missions and the planning. The aim is to find an exploitation rate and identifying all the constraint related to this type of missions. These information helps operators to establish a Post-acquisition Logistics Support system instead of the pre-required one.

4 Conclusion

In this paper, we focus on UAV delivery as a service developed this last decade.

This novelty in an urban area requires that drone delivery operator handles the governance of the UAVs'fleet especially in term of Logistics Support system. The application of Logistics Support for complex system follows a standard the MIL-STD-1388-1A which covers all the steps before the exploitation phase. We speak, in this case, about Integrated Logistics Support.

The nature of the drone delivery operator activity demands operators to adapt the Logistics Support system in function of the exploitation rate and the activity constraint (missions, maintenance, etc.).

As a consequence, we propose a new Logistics Support approach to cover the exploitation phase as a first step after the acquisition in the life cycle of UAVs'fleet.

The Post-Production Logistics Support Analysis (P-P.L.S.A) is inspired by the previous standard and extends same tasks in the exploitation phase.We introduce as a part of the P-P.L.S.A the maintenance organization of the drone delivery operator. In addition, we propose a new task: the activity analysis (502). In this task, we highlight that the exploitation rate of the fleet could impact the Logistics Support System. That is why we should evaluate the Logistics Support requirements continuously.

Also, we want to more elaborate the task 502 in the context of drone delivery operator and evaluate the exploitation rate of the fleet to size the Logistics Support system. For this reason, we can propose a VRP model as a tool of fleet sizing and study the impact of missions in the charging planning or spare parts. These points will be detailed in further works. As a perspective for future work, we will also extend our analysis to the disposal phase.

References

1. Projet de drone: le terminal de livraison (2015). https://www.geopostgroup.com/.fr/actualites/projet-de-drone-le-terminal-de-livraison
2. Bardou, L.: Soutien logistique intégré. Techniques de l'ingénieur. L'Entreprise industrielle (AG5380), AG5380-1 (2000)
3. Chiesa, S., Fioriti, M.: UAV logistic support definition. In: Valavanis, K.P., Vachtsevanos, G.J. (eds.) Handbook of Unmanned Aerial Vehicles, pp. 2565–2600. Springer, Dordrecht (2015)
4. Daum, E.: Le MCO aeronautique: un enjeu pour la coherence capacitaire des armees (2015)
5. University of Defense: MIL-STD-13881B LSA
6. DeGarmo, M.T.: Issues concerning integration of unmanned aerial vehicles in civil airspace. In: The MITRE Corporation Center for Advanced Aviation System Development (2004)
7. Díaz, V.G., Márquez, A.C., Pérès, F., De Minicis, M., Tronci, M.: Logistic support for the improvement of the warranty management. In: Advances in Safety, Reliability and Risk Management, pp. 2813–2820 (2012)
8. Ducret, R.: Parcel deliveries and urban logistics: changes and challenges in the courier express and parcel sector in europe-the french case. Res. Transp. Bus. Manag. **11**, 15–22 (2014)
9. The French Ministry of Ecology, Sustainable Development: T., Housing: Arrete du 11 avril 2012 relatif à la conception des aéronefs civils qui circulent sans aucunepersonne à bord, aux conditions de leur emploi et sur les capacités requises des personnes qui les utilisent.avril 2012 (2012). http://www.legifrance.gouv.fr/affichTexte.do;jsessionid=cidTexte=JORFTEXT000025834953&dateTexte=&oldAction=rechJO&categorieLien=id
10. of system Engineering, T.F.A.: Logistic support. http://www.afis.fr/nm-is/Pages/SoutienLogistique.aspx
11. Feynman, R.: Integrated logistic support. In: Reliability, Maintenance and Logistic Support, pp. 345–375. Springer (2000)

12. Karaağaç, C., Pakfiliz, A.G., Quagliotti, F., Alemdaroglu, N.: UAV logistics for life-cycle management. In: Valavanis, K.P., Vachtsevanos, G.J. (eds.) Handbook of Unmanned Aerial Vehicles, pp. 2601–2635. Springer, Dordrecht (2015)
13. Maddalon, J.M., Hayhurst, K.J., Koppen, D.M., Upchurch, J.M., Morris, A.T., Verstynen, H.A.: Perspectives on unmanned aircraft classification for civil airworthiness standards. Technical report Citeseer (2013)
14. Mueller, T.: On the birth of micro air vehicles. Int. J. Micro Air Veh. **1**(1), 1–12 (2009)
15. Nam, T., Pardo, T.A.: Conceptualizing smart city with dimensions of technology, people, and institutions. In: Proceedings of the 12th Annual International Digital Government Research Conference: Digital Government Innovation in Challenging Times, pp. 282–291. ACM (2011)
16. Omidshafiei, S., Agha-mohammadi, A.a., Amato, C., Liu, S.Y., How, J.P., Vian, J.: Health-aware multi-uav planning using decentralized partially observable semi-markov decision processes. In: AIAA Infotech@ Aerospace, p. 1407 (2016)
17. Ponza, A.: Optimization of drone-assisted parcel delivery (2016)
18. Shukla, S.K., Kumar, S., Selvaraj, P., Rao, V.S.: Integrated logistics system for indigenous fighter aircraft development program. Procedia Eng. **97**, 2238–2247 (2014)
19. Stewart, J.: Google tests drone deliveries in project wing trials. BBC World Service Radio (2014)
20. Washburn, D., Sindhu, U., Balaouras, S., Dines, R.A., Hayes, N., Nelson, L.E.: Helping cios understand "smart city" initiatives. Growth **17**(2), 1–17 (2009)
21. Welch, A.: A cost-benefit analysis of amazon prime air (2015)
22. Williams, K.W.: A summary of unmanned aircraft accident/incident data: human factors implications. Technical report DTIC Document (2004)

Policy Recommendations Supporting Smart City Strategies: Towards a New Methodological Tool

Nils Walravens[✉] and Pieter Ballon

imec-SMIT, Vrije Universiteit Brussel, Pleinlaan 9, 1050 Brussels, Belgium
{nils.walravens,pieter.ballon}@vub.ac.be

Abstract. One of the biggest challenges for cities today is positioning themselves in relation to the debate surrounding the "Smart City" concept. Based on
a thorough value network analysis of 37 international Smart City services in a
doctoral study, a number of policy recommendations are formulated. These
recommendations lead to a new methodology that local governments can use to
build a vision on their Smart City principles and priorities.

Keywords: Smart city · Local innovation · Vision building · Policy
recommendations

1 Introduction

In an era in which technology and urban life increasingly meet and clash in the Smart
City concept, mobile city applications currently are the most concrete proxy to this
interaction between the virtual and physical urban space [1, 2]. Apart from the multi-
billion-dollar economic activity soft- and hardware companies have built around these
services, there is also a large potential for societal impact and public value creation from
mobile apps [2]. However, the nature of this industry makes developing a successful
strategy extremely challenging for cities large and small. The goal of this paper is
formulating concrete policy recommendation targeted at cities confronted with this
challenge. The research question is: "Which inhibiting and contributing factors can be
identified in devising an urban mobile app strategy and what are concrete policy actions
a local city government can take?" [3]. From these recommendations and the lessons
learned from mobile apps, we then derive a methodology that can be used by local
governments to consider their vision on Smart Cities in a more general sense.

2 Methodology and Data

The data gathered to answer the research question and formulate the policy recommendations comes from 37 in-depth international case studies of smart city apps and services,
performed during a four-year research project into mobile application business models
in the region of Brussels, Belgium. This project explored the potential roles of the city
as an actor in the mobile services value network. The investigated cases are represented
in the table below (Table 1).

© Springer International Publishing AG 2017
E. Alba et al. (Eds.): Smart-CT 2017, LNCS 10268, pp. 97–106, 2017.
DOI: 10.1007/978-3-319-59513-9_10

Table 1. Overview of cases

International cases	Brussels official applications
NYC 311	Be.Brussels
Fix My Street	Brussels Gardens
Carambla	City of Brussels
PulsePoint	FixMyStreet Brussels
Stad Mechelen	STIB Mobile
I-amsterdam QR Spots	Visit Brussels
App van 't Stad	Recettes4Saisons
London Bike App	
Berlin Neighborhood	
Ghendetta	
Top Brussels Apps Android	Top Brussels Apps iOS
Öffi	Autosalon Brussels 2013
Airport + Flight Tracker Radar	Brussels Lines
Tripwolf	Tell Me Where
Bioscoopagenda	Météo Bruxelles
Stay.com – City Guides and Maps	City 2 Smart Shopping
Cycle Hire Widget Lite	Brussels Airport
Geolover Travel Guide	700 City Maps
GuidePal Travel Guides	Batibouw 2013
PocketGuide audio travel guide	Brussels Transport Map
Subway and Metro Guide	Djump

The case studies were developed based on desk research, policy documents and 32 semi-structured expert interviews with city officials and smart city or mobile app industry experts. Much more detail on the used methodology and the gathered data can be found in [4]. From this input, we formulate concrete policy recommendations that can serve as inspiration to policy makers. Finally, these policy recommendations are translated and operationalised into a methodology that can support cities in formulating a vision on smart cities and the challenges related to this concept.

3   Policy Recommendations

This section will give a brief overview of the ten policy recommendations, formulated based on the thorough business model and public value analysis of the cases mentioned just above.

3.1   Developing a Vision (and Personifying It)

Developing and supporting a vision is crucial when it comes to complex projects involving lots of different stakeholders. Getting different actors with divergent interests working together is only possible if it is clear what the final goal is and what the preferred avenues of getting there are. While having the "right people at the right place and time" is not something one can necessarily easily (re)create, it can be stimulated and fostered around new profiles and positions. A concrete recommendation is to create the position of a Smart City coordinator (or similar). This person should – regardless of background - be able to bridge different and divergent interests at play and establish a neutral ground where various stakeholders can meet, engage and come up with solutions to pressing urban challenges. Such a profile could come from in- or outside of government and would ideally be a neutral person that is not explicitly linked to a political function or competence. In international examples (e.g. Barcelona, Helsinki, Chicago…) the CXOs of these cities are positioned in different departments of the city (the mayor's office, the digital strategy cell, the innovation department, independently at universities and so on), illustrating how a specific approach per city is recommendable, taking into account the particularities of the (socio-economic and political) context and the approach policy wants to take.

3.2   Setting Ambitious (and Measurable) Goals

In the various debates and operationalisations of what the Smart City could be, it becomes apparent that taking a very specific urban challenge and thinking about how technology could alleviate it, is a perhaps pragmatic, but realistic approach. In order to verify whether the use of a certain technology has had any effect, goals or targets need to be specified beforehand. When considering a Smart City vision and dealing with the vast range of problems an urban environment may entail, one should be ambitious. Of course, projects of this scope and scale are often constrained by budgetary considerations or other factors that may hinder its successful adoption. While we certainly would argue for a healthy dose of realism in developing the envisaged solutions, falling into the trap of extreme pragmatism is a real threat. Given the often complex and intertwined nature of the challenges to be tackled, some out-of-the-box thinking is often required, deliberately sidestepping some of the idées fixes that may exist.

In this sense, we stress again it is a good strategy to "pick your battles" and choose a specific challenge to tackle, but bringing together all required stakeholders together to make sure the project succeeds. Particularly when the domain or the challenge is well-defined and clear, it also becomes easier to be more ambitious and set out some realistic, but forward-thinking goals.

3.3   Breaking Barriers Hindering Cooperation

Political structures and the organisation of government with its different jurisdictions cannot be changed overnight and are a political matter. However, the cases that were analysed have shown that within any given context different actors can find agreement and common purpose. In this challenging environment, pragmatic solutions may not be ideal, but can go a long way in beginning to organise change, without needing deep institutional or regulatory reform.

3.4   Tackling Fragmentation in (Open Data) Policies

This aspect relates strongly to how public bodies are organised and whether a common goal or vision is in place. Issues come to the foreground for example in open data, where different public organisations may open up data on different portals, using different standards, while missing the potential of the combination of certain datasets. The fragmentation of all these initiatives is hindering adoption of open data. Based on the discussions we had during the course of this research, our recommendation would not necessarily be to centralise all the hosting and management of the data for example, but to provide at least a single portal location where all of the available open data pertaining to a certain region can be searched, together with one appropriate license. This means developing a URI-strategy (Uniform Resource Identifier) for an entire region, that allows data and datasets to be linked to each other, without the need of centralising them. Having a one-stop-shop or clearly communicated location for all of a city's open data would be a vast improvement over the current fragmentation.

3.5   Linking up to Existing Expertise

The cases show that in some regions and cities, existing knowledge on topics like smart cities and open data are a blind spot, even though a lot of expertise is available in universities and research centres. Maximally leveraging this knowledge is key in efficient smart city advancement. One example where such expertise should be consulted is in open data licensing strategies. Currently, a wide array of often ad hoc written licenses is being used to regulate open data reuse. This may seem innocent at first, but it could seriously hinder reuse: if a developer wants to create a service that accesses data that is opened under two different licenses, this may lead to legal issues. To avoid this, the use of similar or identical licenses in a region or country would be highly recommended, but this means different approaches should be thoroughly studies by diverse experts, before being implemented.

3.6   Understanding the Return of Open Data

Correctly gauging the economic, societal, cultural etc. potential of opening data is difficult. A few years after the first datasets have been opened and the concept is gaining traction, it is only now slowly becoming possible to measure the successes and failures of opening up. Apart from empirical evidence and counting on the success of experiments and individual projects, academia is also gradually beginning to explore how to operationalise all this, translate it into metrics and develop methods to better assess the value of open data and its reuse potential. The open data concepts are mature enough today to push this type of research further and to invest in developing better tools and metrics, both quantitative and qualitative to estimate and measure the potential value(s) of opening up.

A concrete recommendation towards cities would then be to not only open up data, but link this to an integrated approach in which contact with the market and interested reusers of the data is central. Open data as a concept implies that the data-providing organisation is not necessarily aware of what happens with the data or how it is reused. This can be mediated by engaging with interested reusers and being as open as possible in communicating with external parties. In doing so, it will become much easier for local governments (or any other government organisation for that matter) to assess the return on the public investment made to open up. It is another area where the city is well placed to take up the role of a platform and mediate and facilitate these interactions. Understanding reuse is invaluable input in planning future courses pertaining to open data (which data to open first, the types of support to offer, the types of stimulating measures that may be required to generate a certain of reuse and so on). As such, it is a crucial component of generating indirect public value, of which the effects occur more in the long-term.

3.7 Involving and Engaging Citizens

While it was not the core of this research and the main focus was the perspective of the city, citizen engagement has come to the foreground on several occasions. The role of the end user is becoming more important in creating innovative digital services that are sustainable over time. Particularly in a context of public policy, participation of citizens becomes even more pressing an issue as local, regional and national governments explore new ways of involving citizens through digital channels, while citizens themselves become more vocal using social media for example. If well developed, these new ways of interacting with government can be mutually beneficial and increase the efficiency of internal and external processes for a government organisation.

In many cases however, user needs are often overlooked in the development of new services. The services that can become most successful are those co-created with citizens, tested and iterated upon. Our recommendation in this context would then be to always start service development from the perspective of the user or a clearly identified user need. The important lesson is not to be afraid of citizen participation or involvement, but rather stimulating it and learning from end user experiences. Organising or participating in Living Lab experimentation of new applications or service innovations can play a key role in this. By learning end users' needs, building on top of them and iterating application concepts and ideas will make new services more sustainable over time.

3.8 Joining International Standards and Networks

There are several interesting forms of cooperation and exchange developing on the European and global level. Joining international consortia can be time-consuming and require follow-up, but they can be very valuable towards acquiring best practices and learning from the examples of other cities in a more profound way. Additionally, these networks can be a means to position a city as a Smart City internationally. Co-defining or implementing common and open standards can be beneficial both in the creation of short and long term public value, but it will certainly have an impact on the sustainability of smart city initiatives vying for the latter.

3.9 Fine-Tuning Infrastructure Plans

The main recommendations pertaining to infrastructure projects also relate to the development of a vision. Before large or important infrastructural works are undertaken, they need to be carefully considered and fit the challenge identified. This was the case for network-related projects for a long time (e.g. WiFi coverage in cities), but will remain the case in future IoT-deployments (e.g. sensor networks). Depending on the use case, sensors are being placed in the public space to gather all kinds of data pertaining to life and activity in the city. Anything from noise, pollution, general air quality to crowd movement, traffic and so on can be monitored using various forms of sensor networks. Sensors linked to smartphones can also have a diverse set of applications and services be built on top of them, for example in the areas of tourism, shopping or other experiences. In order to facilitate this, cities are likely to also work with private partners and, similarly to the infrastructure projects described above, need to be careful in their public procurement choices, to guard that the solution selected is not per se the cheapest, but offers the most flexibility and is best geared towards tackling the identified challenges. Since publicly-owned property such as street furniture, roads or government-owned buildings are often involved in these use cases, keeping control over these processes as a public organisation will need to be closely monitored. As some of these sensor network technologies are still farther from market than for example WiFi, including the research community in these tests and eventual rollouts is also recommended.

3.10 Investigating Innovative Funding Concepts

Cities should explore more innovative procurement models such as pre-commercial procurement. In a pre-commercial procurement procedure, the government does not define a fixed list of criteria a solution needs to comply with, but rather defines an overarching "challenge" that needs to be addressed by the developed technology. This solution is not close to market and so what government is actually procuring is the research and development required to develop the tools that are needed to tackle the challenge. At the end of a pre-commercial procurement procedure, the government is then still not required to purchase the solution developed by the tenderer. As the technology can then be considered ready for market, a normal procurement procedure that includes other partners can still be initiated.

The main advantage of this type of approach is that the city is involved in the ideation around new technologies from a very early stage. Without defining exactly what a solution should do, the city can indicate where some core challenges lie for the local government and award companies (local or international, large or small) that can develop potential solutions. Cities do not need to worry about developing technologies in-house, but still ensure that their long-term goals are being addressed, while supporting the commercial sector in doing so. While certainly not an easy model, pre-commercial procurement establishes a new way for public and private to work together to solve long-term goals that benefit citizens. As such, it is highly recommended to explore these new types of cooperation models.

4   Vision Matrix

As far as putting these recommendations into practice goes, the diverse avenues to go down in part depend on political will and creating an understanding of the potential value with policy and decision makers. Practical implementation furthermore and in first instance depends on the vision that needs to be developed, as approaches might differ depending on the domain that is being tackled. However, these recommendations offer a starting point, a set of basic actions that can serve policy makers faced with these complex challenges. The recommendations should be seen as the start of a process, not a finite list of boxes to check towards achieving a next state. It will be a process of interaction and deliberation, not of straight implementation. The role of the policy maker in this process is well put by Robert Reich (cited in [5]) who states that the policy maker is "in a deliberative relationship" and that "rather than making 'decisions' and then 'implementing' them, your role is to manage an ongoing process of public deliberation and education" [5].

In order to enable policy makers to undertake this exercise, this concluding section proposes a methodology. This is derived from the policy recommendations listed above and starts from the premise that a clear vision is the only aspect where everything can start from, which was highlighted in this section as well. But how can such a vision, which is carried throughout the organisation, no matter how big or small, be established? To enable this, we mainly see three areas that are key, based on the idea of the city as a local innovation platform: aspects *internal* to the city as an organisation, those that are *external* and those that are *technological* in nature. We then make abstraction of the policy recommendations cited above and bundle them into terms that are easily understandable, while all relating to the three key areas just mentioned. This results in the following matrix (Fig. 1).

	Vision	Openness	Organisation	Return	Sustainability	Arguments
Internal						
External						
Technological						

Fig. 1. Vision matrix

The goal of this matrix is to support policy makers inventory various important aspects of a vision on the Smart City. It can be used as a methodological tool to develop a carried vision within the city, by using it as a guide in interviews with city managers from diverse departments, as well as public servants working on related problems in their daily practice. The matrix starts from the perspective of the city and asks questions related to internal aspects (the city as an organisation of organisations), external aspects (the relationship with the market, civil society, citizens and so on) and technological aspects (which technologies should be used and to which ends). For each of these, concrete questions are asked related to the six elements in the top row.

- *Vision* comes from the recommendations on developing a vision, setting measurable goals and infrastructure. Questions can include:
  - Internal: What should the focal points of a Smart City be?
  - External: How do we see our position as a city in relation to the market?
  - Technological: Should we be on the bleeding edge of technological development or not?
- *Openness* refers to tackling fragmentation in open data and linking up to data expertise. Indicative questions are:
  - Internal: How should open data be used within our organisation and which department takes final responsibility?
  - External: What do our open data and data sharing policies look like? What are our guiding principles in opening data to external actors?
  - Technological: Which (open) technologies do we use in the back end?
- *Organisation* refers to breaking barriers for cooperation and involving citizens. Questions include:
  - Internal: How do we organise ourselves internally around the Smart City question and where do responsibilities lie?
  - External: (How) do we organise structural and systematised engagement with the market, civil society, citizens and so on?
  - Technological: Which processes need to be put in place around ICTs and their procurement? Which safeguards against (vendor) lock-in do we need?
- *Return* links to the recommendations of identifying the return of open data and questions can be:
  - Internal: What kind of return do we expect from certain initiatives (e.g. opening up data)?
  - External: What do we expect from the market or other external stakeholders in return for certain support mechanisms (e.g. open data take-up supporting initiatives)?
  - Technological: Which return do we expect from certain technological choices (e.g. open service standards)? What do we expect out of service contracts?
- *Sustainability* refers to joining networks and standards, and applying innovative funding concepts, with questions such as:
  - Internal: How can (open) standards be adopted to ensure the longevity and compatibility of our systems with others?
  - External: What can we learn from external actors via thematic networks or standardisation organisations? What type of financial support can we offer external actors that is sufficiently sustainable?
  - Technological: Which technologies do we choose so that are sustainable over longer periods of time, without creating (vendor) lock-in or compatibility issues?

Arguments finally, is a more transversal category that can be used to capture common arguments for or against developing or implementing a certain vision. This could for example be the case with a concept such as open data, which is still being contested by some policy makers or governmental departments. This category allows capturing these arguments, identifying which ones return and may need to be tackled before a carried vision can be developed.

The matrix is used to structure conversations with policy makers and public service practitioners. Once a series of interviews has been conducted, the different cells can be easily compared to one another, allowing the identification of commonalities and differences. The common points can be included in a vision, while the differences can be further discussed, e.g. in a workshop or co-creation setting, until a minimal consensus can be achieved. From all of this input, a carried vision on Smart City challenges can be derived and the role of the city as a local innovation platform should become much more tangible.

5   Contribution and Limitations

The contribution of this paper mainly lies in the concrete recommendations made towards developing a thought-through smart city strategy or policy. Many urban areas today share similar challenges when it comes to incorporating new technologies: today this has been shown to mainly be an issue of organization and governance, more than anything else. The policy recommendations outlined by the paper offer new insights as to how these organizational challenges can be tackled in the volatile area of smart cities. The proposed methodology has been used successfully in three projects, involving a city government, a large government administration and a publicly funded research organisation.

The method still needs to be refined further and placed in the framework of a more holistic and all-encompassing methodology. Furthermore, the "Arguments" category is an odd-one out and could perhaps better be treated outside of the "Internal, external, technological" structure.

6   Conclusion

Based on the value network analysis of 37 international Smart City services, this paper presents ten policy recommendations to local governments that are struggling to position themselves in the context of current Smart City debates. Based on these recommendations, the vision matrix is assembled as part of a methodology to support cities in formulating a vision on the topic. By applying this matrix in a co-creation setting with different city departments, policy makers and practitioners, a vision that is carried throughout the organisation can be composed. This method has been successfully applied in three projects with regional and local government and will be further refined in future projects. The goal is to establish a robust tool and methodological approach that can help cities in tackling the complex challenges facing the Smart City.

References

1. Townsend, A.: Smart Cities. Norton & Company, New York (2013)
2. Walravens, N.: Qualitative indicators for smart city business models: the case of mobile services and applications. Telecommun. Policy **39**, 218–240 (2015)

3. Walravens, N.: Should there be an app for that? inhibiting and contributing factors to the development of a mobile smart city strategy for brussels. Brussel Stud. **88**, 1–11 (2015)
4. Walravens, N.: Should there be an app for that? public value creation from 'smart' mobile application initiatives for brussels and local government, Ph.D Thesis, VrijeUniversiteit Brussel, 509 p. (2016)
5. Bozeman, B.: Public-value failure: when efficient markets may not do. Publ. Adm. Rev. **62**(2), 145–161 (2002)

Predicting Car Park Occupancy Rates
in Smart Cities

Daniel H. Stolfi[1]([✉]), Enrique Alba[1], and Xin Yao[2]

[1] Departamento de Lenguajes y Ciencias de la Computación,
University of Malaga, Malaga, Spain
{dhstolfi,eat}@lcc.uma.es

[2] CERCIA, School of Computer Science, University of Birmingham, Birmingham, UK
x.yao@cs.bham.ac.uk

Abstract. In this article we address the study of parking occupancy data published by the Birmingham city council with the aim of testing several prediction strategies (polynomial fitting, Fourier series, k-means clustering, and time series) and analyzing their results. We have used cross validation to train the predictors and then tested them on unseen occupancy data. Additionally, we present a web page prototype to visualize the current and historical parking data on a map, allowing users to consult the occupancy rate forecast to satisfy their parking needs up to one day in advance. We think that the combination of accurate intelligent techniques plus final user services for citizens is the direction to follow for knowledge-based real smart cities.

Keywords: Smart city · Smart mobility · Parking · K-means · Time series · Machine learning

1 Introduction and Related Works

Finding an available parking space is hard in most big cities, especially in the city center. Off-street car parks are a viable alternative, especially when the number of inhabitants in urban areas is increasing and expected to rise to 75% of the world's population by 2050 [1]. On-street parking spaces are quite limited and usually it is cheaper to find an off-street car park or pay and display bays rather than wasting time (and fuel) in finding a free space. Not to mention the health consequences [2] provoked by an increase of not only air pollution but also drivers' stress. However, even paid spaces are scarce nowadays as city infrastructures have not grown in line with population growth.

Fortunately, smart city initiatives are changing this [3]. One of the main aspects of a smart city is the so-called Internet of Things (IoT). The main idea is to know the state of a city by using sensors to monitor such data as the road traffic state, temperatures, pollution levels, and car parks' occupancy rates. Although monitoring single parking spaces is not economically viable, it is possible to count the number of vehicles entering and leaving an off-street

© Springer International Publishing AG 2017
E. Alba et al. (Eds.): Smart-CT 2017, LNCS 10268, pp. 107–117, 2017.
DOI: 10.1007/978-3-319-59513-9_11

car park and make these data publicly available to help make decisions (and predictions) based on them.

The prediction of car park availability is the subject that has been studied in a context of smart cities, especially now when most parking facilities have installed sensors as part of their infrastructure.

In [4], the authors fit a continuous-time Markov model to predict future occupancies in several parking locations to propose different alternatives to drivers. They consider not only the car park occupancy rate but also the estimated time of arrival obtained from the vehicle's navigation system in which the calculations are done. They provide two *ad hoc* examples to test their proposal, showing promising results. In this article we take a different approach where, instead of using a navigator, any Internet capable device, such as a mobile phone, can be used to check the current/future state of the desired car park.

Two smart car park scenarios based on real-time information are presented in [5]. The authors use historical data made available by the authorities of the cities of San Francisco, USA and Melbourne, Australia. They employ Regression Tree, Neural Networks and Support Vector Regression as prediction mechanisms for the parking occupancy rate. Their experiments reveal that the regression tree using the historical data in combination with times and weekdays, performs best for predicting parking availability on both data sets. We have analyzed different predictors in our analysis, however, it would be interesting to compare our results to those produced by these alternative predictors in the future.

In [6] a methodology for predicting parking space availability in Intelligent Parking Reservation architectures is proposed. It consists of a real-time availability forecast algorithm which evaluates each parking request and uses an aggregated approach to iteratively allocate parking requests according to drivers' preferences, and parking availability. They employ historical information of entering and leaving to update and predict the availability for each parking alternative. The results provided, obtained from contrasting predictions with real data, show that the forecast is adequate for potential distribution in real-time. Our approach differs from this proposal in that we study different predictors and do not interact with the current demand, relying just on the historical data.

In short, our proposal consists in studying the different prediction strategies to analyze the historical occupancy rates of car parks and forecast the future availability, presenting this information to the users in a web page.

The rest of this paper is organized as follows: Sect. 2 describes the system architecture, including the web page and the predictors. In Sect. 3 we discuss the predictor techniques, the training and testing stages, and our results. Finally, in Sect. 4, conclusions and future work are presented.

2   System Architecture

The architecture of our system (Fig. 1) comprises Downloaders, the Data Parser, the Database, the Predictor and the Web Page in which both, the current state of the car parks and the predictions made, are presented to the public.

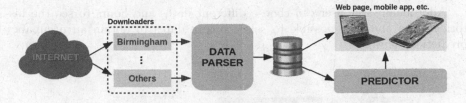

Fig. 1. Schema of the data parser.

2.1 Downloaders

These modules access different data sources available from the Internet to obtain the occupancy data of the car parks defined in the system. They are set up with the source URL, the frequency of readings, as this has to be adapted according to each data source, and the possible data transformations (CSV, XML, etc.) to be completed before feeding the Data Parser. Note that in this study we are working with just one data source.

2.2 Data Parser

The Data Parser processes the data provided by the Downloaders and stores them in the Database. It also checks the validity of the car parks, creating new ones if necessary, whilst avoiding data redundancies.

2.3 Database

The Database stores the data collected from each car park so that it can be shown on the web page. We store the code, description, capacity, latitude and longitude of each car park, as well as the city to which it belongs. Periodically, we also store occupancy data for car parks consisting of spaces used, state, and last updated time.

Additionally, the historical data of the car parks is obtained by the Predictor from the database to be used for forecasting their future occupancy.

2.4 Web Prototype

Data stored in the database can be shown at any time in our web page and mobile app. We present each car park geolocated in its real geographical position in the map by using the library Leaflet[1] and the tiles from OpenStreetMap[2].

Figure 2 shows snapshots of the web page as visualized on a computer desktop and on a mobile phone. We can see that each car park is shown as a circle whose size is proportional to the number of parking spaces and its color represents the occupancy rate as shown in the upper scale, i.e. blue for totally free and red for full. Car parks whose data is out of date are shown in black.

---

[1] http://leafletjs.com/.
[2] http://www.openstreetmap.org/.

Additionally, the user can choose different dates and hours to see the historical data and it is possible to select future dates to obtain an occupancy prediction, as well.

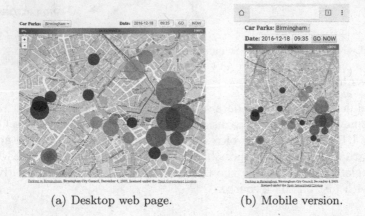

(a) Desktop web page.   (b) Mobile version.

Fig. 2. Web page and mobile prototypes presenting the geolocation, state, capacity, and occupancy of each car park. (Color figure online)

2.5 Predictor

Data stored in the database is also used to predict future occupancy of the car parks. We have experimented with six different predictors (Fig. 3): Polynomial Fitting [7], Fourier Series [8], K-Means [9], KM-Polynomials, Shift & Phase, and Time Series [10] which are all described in the next section. We have selected them for this initial study because they are simple, easy to implement, and they allow us to represent each car park with just a few parameters. Furthermore, they are present in the open data provided by cities nowadays.

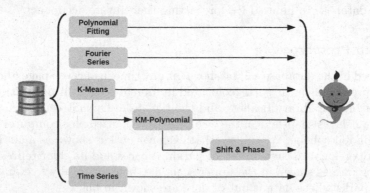

Fig. 3. Different predictors tested and the existing relationships between some of them.

3   Prediction Techniques

We wish to address the prediction of the future occupancy rates of car parks in a city. For our prototype we have chosen the data set "Parking in Birmingham" published by Birmingham City Council[3] in the United Kingdom, licensed under the Open Government License v3.0. It includes the car parks operated by NCP (National Car Parks) in that city, and is updated every 30 min from 8 AM to 5 PM. In our study, we worked with data from Oct. 4th 2016 to Dec 19th 2016 (11 weeks).

The data provided is not very accurate as sometimes the sensors are faulty or even, the whole data set may not be updated for a whole day. To address these situations we implemented a filtering stage before the data processing as follows:

1. The occupancy rate is calculated based on each car park's capacity.
2. The percentage values beyond the range (0–100%) are adjusted to these limits.
3. Out of date data is discarded.
4. If the variability of a car park's occupancy for an entire day is below 5% it is assumed that it is due to faulty sensors and that day is discarded.
5. Data on a car park that is below 5% for an entire day is also discarded.
6. Car parks without data are excluded from the study.

Figure 4(a) shows the occupancy data available for all the car parks and dates after filtering the initial data set and in Fig. 4(b) the data distribution of the car park occupancy on weekdays is depicted as a boxplot.

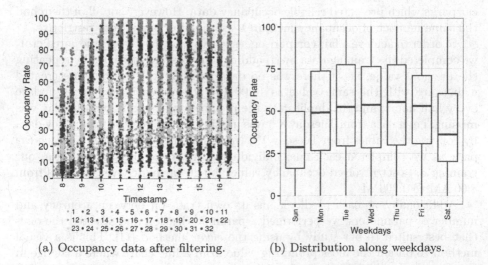

(a) Occupancy data after filtering.   (b) Distribution along weekdays.

Fig. 4. Occupancy data from the 32 car parks and their distribution on weekdays.

---

[3] https://data.birmingham.gov.uk/dataset/birmingham-parking.

We can see that occupancy rates decreases on Saturdays and Sundays as expected, while they are quite similar throughout the rest of the week. All in all, we finally had a working data set consisting of 32 car parks and 36,285 occupancy measures.

Throughout our study we used the Mean Squared Error (MSE) to test the quality of the predictions made, not only in the training stage but also in the testing stage. Equation 1 presents the MSE formula where y_i are the measured real values, f_i are the fitted ones, and n is the number of observations.

$$MSE = \frac{1}{n} \sum_i (y_i - f_i)^2 \tag{1}$$

3.1   Training

Prior to training, data ought to be processed so as to guarantee a fair comparison between the different predictors used. We wished to predict the occupancy rate of each car park over an entire week, consequently, we decided to use a different predictor for each car park and weekday to conduct this first initial study. We selected the first ten weeks of data (Oct 4th to Dec 12th) for training and left the eleventh week for the testing stage to simulate the real use of the web page by a user.

As we have pointed out sometimes sensors fail. To address this, first, we discarded data from a car park for an entire day when more than 25% of the measures were missing. Second, if a car park did not have at least one weekday of training data, it was also excluded.

After applying this second filter to the training data set, we finished with 29 car parks which presented reliable occupancy data. However, not all of them had the same number of occupancy measures, as our filter is not very restrictive.

In order to achieve a fair comparison, especially for the Time Series predictor, we completed the training data by (i) adding non-existent measures by repeating the previous value, i.e. if there was no data at 11:30 AM we therefore created a measure with the same value as 11:00 AM (Birmingham car parks update every half hour); and (ii) duplicating the previous weekday if an entire day was missing, i.e. if data from Tuesday 8th was missing we created the occupancy data by copying the values from Tuesday 1st. Note that this was checked for each car park as we completed each one, individually. After completing the data, our training data set involved occupancy values for 29 car parks over ten days from 8:00 AM to 4:30 PM.

Additionally, as each predictor has its own trade-off between accuracy and number of parameters, we performed a parameterization and selected the ones that best suited to our study by using the *elbow* method [11]. This is a visual method to obtain the most promising value from a line chart where a change in the slope looks like the elbow of an arm.

To improve the training process we decided to use K-Folds cross validation. We used 10 sets (we have 10 weeks of data for training) where each training set consisted of 32,886 occupancy data values (29 car parks, 9 weeks, 7 days, 18 values per day). To obtain the average MSE values we tested each predictor on the remaining week (3,654 values). By selecting a different test week we obtained ten different training and testing sets to train our predictors as discussed below.

Polynomial Fitting (P). This predictor consists in a polynomial fitted to each car park and weekday. We studied different polynomial degrees to find which value presented a low average MSE. We also wished to keep a reduced number of parameters to represent each car park and weekday. Figure 5(a) presents the MSE values obtained after using cross-validation for all ten training processes for Birmingham. We can see that according to the *elbow* method, polynomials of second degree are the best choice to be used in this predictor for the ten cases because they have only a few parameters and good precision.

Fourier Series. This predictor consists in fitting a Fourier series to each car park and weekday. In this case we considered different numbers of components. Since they are composed of pairs of sines and cosines, the different alternatives tested are always odd numbers (a constant component is included, as well). In Fig. 5(b) the MSE values obtained after using cross-validation to train the Fourier predictor are depicted. The *elbow* method clearly states that using just three components (a constant, a sine, and a cosine) is the best choice in all cases, not only because of the change in the slope, but also because it leads to low MSE values without increasing the number of components.

***K*-Means.** Clustering by using *K*-Means is a method that allows grouping pairs of car parks and weekdays in different clusters whose centroid represents the whole set of occupancy measures in the group. It is an interesting way of describing a set of car parks which behave similarly. We tested up to ten clusters to decide which option was better according to the MSE values and the *elbow* method. Figure 5(c) presents the MSE values obtained for each fold and number of clusters. It can be seen that three clusters is a good choice for this predictor for all the training folds.

KM-Polynomials. This predictor fits a polynomial to the existing centroid points of each cluster calculated by *K*-Means. This step is necessary to improve the accuracy of the predictions by interpolating a polynomial to the points in each centroid as they are spaced according to the frequency of the measures, i.e. 30 min. In Fig. 5(d) the MSE values obtained for different polynomial degrees are depicted. We can see that, using the *elbow* method, the best degree of the polynomials matches the one selected for the Polynomial Fitting predictor. This was somewhat expected, as the centroids ought to represent a set of measures which are the same as the ones used to obtain the aforementioned polynomials.

Shift & Phase (SP). To improve the accuracy of the prediction even further, we defined a new predictor which uses the KM-Polynomials calculated in the previous section and adds two new parameters (δ and ϕ) in order to modify the shift (y axis) and the phase (x axis) of the original polynomial as shown in Eq. 2.

$$F(x) = (a_0 + \delta) + a_1 \times (x + \phi) + a_2 \times (x + \phi)^2 + \ldots + a_n \times (x + \phi)^n \qquad (2)$$

Then, by using a weighted nonlinear least-squares estimation [12], the values for δ and ϕ for each pair of car park and weekday were obtained, so that a car park's occupancy rate could be predicted by following the process shown in Algorithm 1.

Algorithm 1. Occupancy Prediction.

---

   function OCCUPANCYPREDICTION(car_park, wd, $time$)
      $cluster\_id \leftarrow getClusterId(car\_park, wd)$
      $coefs \leftarrow getPolynomialFitting(cluster\_id)$
      $(\delta, \phi) \leftarrow getShiftPhase(car\_park, wd)$
      $occupancy \leftarrow getOccupancy(time, coefs, \delta, \phi)$
      return $occupancy$
   end function

The main function receives as parameters the car_park identity, the weekday (wd), and the $time$ at which we want to know the occupancy. Inside the function, the corresponding $cluster\_id$ is obtained based on the car_park and the weekday wd, as the first step. Second, the coefficients $coefs$ of the polynomial fitted to the cluster's centroid are obtained. Third, δ and ϕ for the car park and weekday are also obtained. Finally, the $occupancy$ value is calculated by using the formula in Eq. 2 whereas x is the $time$ parameter.

Figure 5(e) shows the best data set according to the MSE values obtained when training SP. After all these experiments, it is clear that data set 1 (when we train with 2 to 9 and test on 1) presented the best results, i.e. the lower average MSE values for all the predictors trained by using cross validation.

Time Series (TS). To train the Time Series (TS) predictor a different approach was followed, as a consecutive, ordered number of time periods (weekdays) are needed, which makes it impossible to use k-folds. We therefore trained a different time series for each car park and weekday to be consistent with the other predictors analyzed here.

Figure 5(f) shows the MSE values obtained when training the TS predictor with different numbers of weeks. It is worth noting that having more data did not imply computing the best prediction according to our experiments. Nevertheless, it was something to be analyzed at a later date, as there were not enough data to make a solid conclusion at that point. Furthermore, a variation in the test week such as a bank holiday may also increase the MSE value.

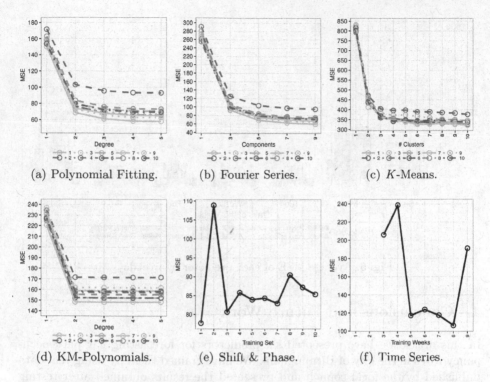

(a) Polynomial Fitting.   (b) Fourier Series.   (c) K-Means.

(d) KM-Polynomials.   (e) Shift & Phase.   (f) Time Series.

Fig. 5. Average MSE values obtained when training the different predictors by using k-fold cross validation (10 folds). (a)–(d) also shows the parameterization performed.

3.2 Prediction

In this next step we compare the predictions made by the predictors trained in the previous sections. To do so we predicted the occupancy rates for seven unseen days (from Dec 13th to Dec 19th) and compared the MSE values obtained for each region. We did not complete this data set as we did in the training stage as isolated values are also useful to test our predictions providing they are produced by reliable sensors. All in all, we have tested our predictor on 3,425 occupancy values: 480 for Sunday, 468 for Monday, 493 for Tuesday, 493 for Wednesday, 501 for Thursday, 522 for Friday, and 468 for Saturday.

Figure 6 shows the boxplots of the distribution of the MSE values for each predictor. We can see that TS performed best for weekdays followed by P and SP which show very good results for the whole week except Mondays. KM and KP are the worst predictors as they are based on just the three clusters defined in training. Fourier improved upon KM and KP on all weekdays except on Mondays which was clearly the hardest day for the predictors, followed by the weekend. Saturdays and Sundays are the days when people do not follow a clear pattern of behavior as they do on working days, which, in part, explains the larger MSE values observed. On the other hand, the occurrence of large MSE values observed for Monday has to be further investigated as they clearly differ from other days.

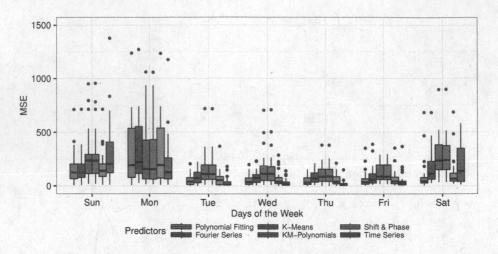

Fig. 6. Average MSE of each predictor by weekdays.

4   Conclusions and Future Work

In this article we have presented six predictors for forecasting car park occupancy rates in the city of Birmingham. We have trained them by using real data published by the local council and presented the results obtained after testing them with one week of unseen parking data.

Although there is no clear winner, the time series predictor seems to show the best results. Shift & Phase also has good results, especially if we take into account that it is simpler and requires just five parameters to predict a car park's occupancy rate instead of a series of values.

Our proposal is a novel service as although there are web pages offering information on car park's occupancy rates, they rarely make predictions of the next day's state.

As a matter for future work we wish to develop a mobile app, repeat this study using a larger training data set, and include new predictors in the comparison, e.g. a multivariate predictor. Additionally, we wish to address another method for missing values such as average, previous days.

Acknowledgements. This research is partially funded by the Spanish MINECO project TIN2014-57341-R (http://moveon.lcc.uma.es). Daniel H. Stolfi is supported by a FPU grant (FPU13/00954) from the Spanish Ministry of Education, Culture and Sports. University of Malaga. International Campus of Excellence Andalucia TECH.

References

1. Bakici, T., Almirall, E., Wareham, J.: A smart city initiative: the case of barcelona. J. Knowl. Econ. **4**(2), 135–148 (2013)
2. Hertel, O., Jensen, S.S., Hvidberg, M., Ketzel, M., Berkowicz, R., Palmgren, F., Wåhlin, P., Glasius, M., Loft, S., Vinzents, P., Raaschou-Nielsen, O., Sørensen, M., Bak, H.: Assessing the impacts of traffic air pollution on human exposure and health. In: Jensen-Butler, C., Sloth, B., Larsen, M.M., Madsen, B., Nielsen, O.A. (eds.) Road Pricing, the Economy and the Environment. Advances in Spatial Science, pp. 277–299. Springer, Heidelberg (2008)
3. Neirotti, P., De Marco, A., Cagliano, A.C., Mangano, G., Scorrano, F.: Current trends in Smart City initiatives: some stylised facts. Cities **38**, 25–36 (2014)
4. Klappenecker, A., Lee, H., Welch, J.L.: Finding available parking spaces made easy. Ad Hoc Netw. **12**(1), 243–249 (2014)
5. Zheng, Y., Rajasegarar, S., Leckie, C.: Parking availability prediction for sensor-enabled car parks in smart cities. In: 2015 IEEE Tenth International Conference on Intelligent Sensors, Sensor Networks and Information Processing (ISSNIP), pp. 1–6 (2015)
6. Caicedo, F., Blazquez, C., Miranda, P.: Prediction of parking space availability in real time. Expert Syst. Appl. **39**(8), 7281–7290 (2012)
7. Fan, J., Gijbels, I.: Local Polynomial Modelling and its Applications: Monographs on Statistics and Applied Probability, vol. 66. CRC Press, Boca Raton (1996)
8. Butzer, P.L., Nessel, R.J.: Fourier Analysis and Approximation, vol. 40. Academic Press, New York (2011)
9. Hartigan, J.A., Hartigan, J.A.: Clustering Algorithms, vol. 209. Wiley, New York (1975)
10. Fuller, W.A.: Introduction to Statistical Time Series, vol. 428. Wiley, New York (2009)
11. Sugar, C.A.: Techniques for clustering and classification with applications to medical problems. Ph.D. thesis, Stanford University (1998)
12. Draper, N.R., Smith, H., Pownell, E.: Applied Regression Analysis, vol. 3. Wiley, New York (1966)

Predicting Individual Trip Destinations
with Artificial Potential Fields

Alessandro Zonta[1($\boxtimes$)], S.K. Smit[2], and Evert Haasdijk[1]

[1] Vrije Universiteit Amsterdam, Amsterdam, The Netherlands
{a.zonta,e.haasdijk}@vu.nl
[2] TNO, The Hague, The Netherlands
selmar.smit@tno.nl

Abstract. This paper presents a method to model the intended destination of a subject in real time, based on a trace of position information and prior knowledge of possible destinations. In contrast to most work in this field, it does so without the need for prior analysis of habitual travel patterns. The method models the certainty of each POI by means of a virtual charge, resulting in an artificial potential field that reflects the current estimate of the subject's intentions. The virtual charges are updated as new information about the subject's position arrives. We experimentally compare a number of update rules with various parameter settings, showing that it is important to take the distance to a potential destination into account when updating the charge.

Keywords: Human behavior · Intention analysis · Destination prediction · GPS · Trajectory database

1 Introduction

Monitoring the movement of individuals offers support for smart decision making in, for example, traffic control, crowd monitoring or location-based services. Insight into the current destination of an individual's movements in particular can provide substantial benefits at a central control level as well as at individual level. At an individual level, location-based services could benefit from such knowledge to reserve resources at their arrival, provide relevant reminders on their to-do lists or suggest places to meet each other [1].

Aggregated knowledge of the destination of individuals in transit can for instance help control rooms to adapt the traffic lights around a city dynamically, or schedule the operation of large infrastructural elements such as ferries or moveable bridges so as to maximise the flow of traffic. In a security or military context, information about the intended destination or itinerary of suspicious individuals or opposing units provides obvious benefits for planning safe transport, (troop) deployment and interception with minimal collateral damage.

With the growing availability of aerial imagery, inexpensive GPS tracking and smartphone usage data, the last few years saw a substantial number of studies

© Springer International Publishing AG 2017
E. Alba et al. (Eds.): Smart-CT 2017, LNCS 10268, pp. 118–127, 2017.
DOI: 10.1007/978-3-319-59513-9_12

that focussed on the use of such data to predict trajectories, waypoints and trip destinations e.g., [1–7]. Most work to date analysed data sets of logged trajectories to develop models of habitual movement patterns that can subsequently be employed to predict an individual's movements.

One use of location data is the identification of 'key locations' –destinations or waypoints– that can be used as a basis for predicting an individual's movement. Ashbrook and Starner [1] used clustering approaches to identify key locations from GPS data in an unsupervised manner and subsequently derive a Markov model to predict the probability of transition between these locations. Liao et al. [7] used labelled data to develop a Markov model that extracts and classifies key locations from GPS data. A Bayesian network then models habitual transportation patterns that predict the next location on an individual's path. Scellato et al. [3] identified places that individuals habitually visit based on the length of time a user stayed at a particular position. They then used methods from non-linear time series analysis to predict which of these places an individual is heading for.

A substantial amount of research focusses on habitual traces such as standard paths in an individual's everyday routine. Nicholson and Noble [2] generated a Markov model from smartphone connectivity data that was shown to be capable of accurately predicting individual movement after analysing a week's data. Ziebart et al. [6] analysed data from 25 taxi cab drivers to learn standard routes and developed Markov models to predict the next turn, the itinerary and the destination of a ride. Sadilek and Krumm [5] extracted regularities from each individual's location data and learned their association with particular days of the week. This allowed accurate prediction of the individual's itinerary. Fallis [8] based their predictions of movement on comparisons of an individual's current path with historical traces using a combination of probabilistic inference and path extrapolation.

Lorenzo et al. [9] and Ying et al. [10] used clustering techniques to develop maps that link locations to probable activities. Zheng et al. [11] considered the location data of multiple individuals to identify gatherings and joint movement.

All these approaches have in common that they rely on historical data to identify waypoints and destinations, and to develop models that can eventually predict an individual's destination. Such data is not always available in sufficient quantity, for instance in rural areas, or in one-off situations such as manifestations, festivals or military settings. In this paper, we present a model to predict an individual's destination in real time without the need for prior analysis of habitual travel patterns. The model designates possible destinations ("points of interest" or POIs) as virtual charges that together form an artificial potential field. As location information comes in, the charges are updated so that the POIs that the individual moves towards increase their charge and ones that the individual moves away from decrease. Thus, the charge of the most likely destinations increase, allowing the model to identify an individual's probable target.

Artificial potential fields have been extensively studied as a method in path planning, with obstacles virtually charged so that they repel the subject for obstacle avoidance e.g., [12–14] or targets virtually charged so that they attract the subject e.g., [15,16].

We experimentally validate and analyse the proposed model in two scenarios. The first uses a simulator that is able to realistically simulate the movement of thousands of people in the city of Rijswijk in the Netherlands [17]. The second scenario uses actual GPS trajectory data collected by Microsoft Research Asia of 182 users over a five-year period [18–20].

2   Tracking and Prediction System

The basic idea was to build a model able to follow the intention of a tracked subject toward its destination in an urban context. We assume that potential destinations are known in advance; we label them as POIs. Every POI carries a charge q that indicates how likely the POI is to be the subject's target. This defines an Artificial Potential Field (APF) where there are several points, i.e., the POIs, attracting another point, i.e., the tracked person. This APF serves as a model of the subject's destination and intentions and the highest charged POI represents the most likely estimate of the model for the subject's destination.

Equation (1) is used by the model to compute the attraction of the POIs on the tracked person:

$$|E| = \frac{|q|}{d^2} \ .$$
(1)

The formula corresponds to Coulomb's law equation to compute the magnitude of the electric field E created by a charge q at a certain distance d.

To reflect the movement of the subject and use it to adjust the estimate for its destination, an update rule is defined. The update rules are stated in terms of the angle of vision (FOV), the distance from the subject to the POI, the APF and the angle between the POI and the subject's direction of movement. There are two versions of the update rule: (2) uses the angle and (3) also includes the distance.

$$\Delta c = F \cdot (s_1 e^{\beta w_1})$$
(2)

$$\Delta c = F \cdot (s_1 e^{\beta w_1} + \frac{1}{s_2 d^{w_2}})$$
(3)

with $F = 1$ if the POIs are inside the FOV, otherwise $F = -1$. Variables s_1 and s_2 are used to weight the possible influence of β and d to the final result. w_1 regulates how dependent is Δc to the angle β. Small values reduce the influence of the angle and consequently, all the POIs are updated with the same amount even though they have different directions. The variable w_2 is enhancing the difference between the increment of closer and farther POIs. For POIs within the FOV, as for POI_1 in Fig. 1, the angle with the edge of the FOV (γ_1 in Fig. 1) is used in (2) and (3) to compute Δc. The maximum increase is achieved when a POI is exactly in the direction of the movement. Similarly, for POIs outside the

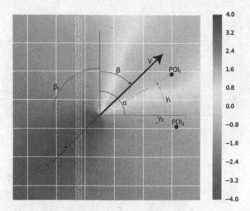

Fig. 1. Representation of virtual charge update. V is the movement direction, α is an algorithm setting and the increment grows following the angles β and β_1. The colours represent the POI's change in charge using (2) with $s_1 = 1$ and $w_1 = 0.01$. γ_1 and γ_2 are used by the model to compute the increment or decrement of the two POIs.

FOV (e.g., POI_2 in Fig. 1) the angle with the edge of the FOV (γ_2 in Fig. 1) is used in (2) and (3) to compute the *decrement* that is maximal for POIs directly opposite the movement direction.

Figure 2 shows how the update rule affects the APF to follow the intention of the subject. Figure 2a, b, c, d, and e represent the evolution of the APF following the trajectory highlight in black, meanwhile Fig. 2f shows the evolution of the POIs' charge.

Two different methods are considered to compute d and γ. The first option is to use the angle and distance according to the Euclidean Distance (using the Haversine formula) and angle representing the path 'as the crow flies'. The second option is to use a path planner to determine the shortest path towards a POI. In that case, γ is defined as the angle towards the first way-point on the path, and the distance is the length of the planned path towards the POI.

Finally, we have tried a variant where a POI's charge is only updated when the APF is not in line with the current direction of movement.

In the following sections, we use the naming convention N for the rules not using the distance and D for the ones using it and specifically DE if using the Euclidean Distance or DP when using a path planner and finally A indicates that is only adjusted when not in line with the current forces.

3   Experimental Settings

To test performance and robustness, we have selected two data-sets of trajectories, several different approaches and varied their parameters. The two data-sets differs from each other on, for example, the length of the trajectories, the area where the POIs could be found and the transportation systems used. The POIs are determined in the same manner for each dataset: the endpoints of all the

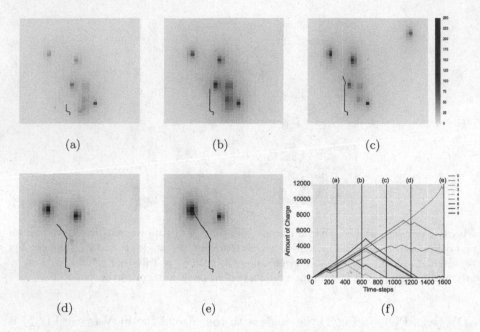

Fig. 2. Representation of how the charges in a Potential Field change. The black line represents a trajectory; the dots describe the intensity of the virtual charge for each POI. (f) shows the time-step where the heat-map are taken, with the charge intensity for each POI.

trajectories in the dataset are labelled as POIs, and a clustering algorithm is used in order to reduce the possible uncertainty of the locations. The objective was to determine if there is a robust setup that works reasonably well on both of the test-cases using the same parameter values since changing them would not be feasible in an operational setting.

Datasets. The first data-set, is created using a simulator of people's movement around a neighbourhood [17]. Each person is generated from open census data and a pattern-of-life, and a social network is created to let behave as normal people with friends and family. The second is a GPS trajectory data-set collected by Microsoft Research Asia [18–20] (Geolife Trajectories) tracing 182 users in a period of five years. The data-set contains a broad range of outdoor movement, including life routine as going home or go to work, entertainment and sports activities such as shopping, sightseeing, hiking and cycling. Cars, bus and taxies trajectories are also present in the data-set.

Figure 3 shows how different the two data-sets are in term of trajectories' length.

Parameters. The model has five parameters: the angle of vision and the parameters in (2) and (3). A parameter sweep was performed to find the configuration

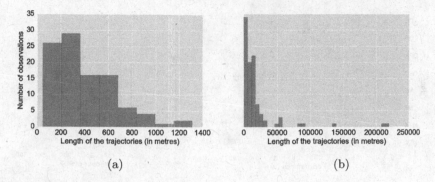

Fig. 3. Length of the two data-sets' trajectories in metres. (a) represent the IDSA trajectories and (b) shows the Geolife Trajectories.

Table 1. Set of parameters tested.

α	w_1	w_2	s_1	s_2
120°	0.005	0.100	0.250	0.100
180°	0.010	0.250	0.500	0.250
240°	0.020	0.500	0.100	0.500

able to achieve the best prediction. In the following paragraphs, we will use *setting* as a set of parameter. Table 1 shows the parameter tested.

Evaluations. The performance is defined as the percentage of the trip length with the target being the highest charged POI in the trajectory. 100 people were tracked with the same setting and the average performance of all of them is considered as the performance of that settings. We use a pairwise non-parametric comparison based on the Wilcoxon-Mann-Whitney test to statistically assess differences in rule performance. We used a Simple Linear Regression Analysis to compute which parameters have most influence on the result. The same trajectories are simulated with three sets of POIs to evaluate how the better rule behaves: firstly, for each subject, a set containing the destinations of all 100 people –the *normal set*– is used. A smaller set is created by removing 50 randomly selected POIs, taking care not to remove the actual target, and a larger set is created by adding 200 further destinations from subjects that are not in the current sample.

4 Results and Discussion

The median and the interquartile range of the performances per update rule are shown in Fig. 4a. Rules based upon the Euclidean distance(DE) achieved the best performance on both data-sets both on median ($p < .001$) and maximum.

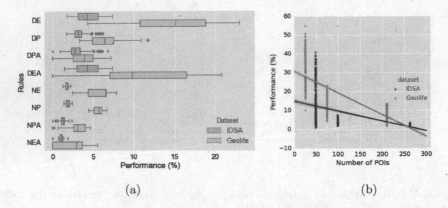

(a)   (b)

Fig. 4. (a) shows the update rules' performances. The performance is defined as the percentage of the number of time-steps with the target being the highest charged POI in the trajectory. Every bar shows the median and the interquartile range over the median of 100 people tracked per setting. The D on the name correspond to Distance with E for the Euclidean version and P for Path version. A means the usage of the APF and N the angle without the distance. (b) shows the results with different number of POIs with the DE. From the normal set of POIs, some of them are deleted to simulate fewer POIs or different random ones are added to simulate more POIs.

On the Geolife data-set, these rules yielded into a performance of over 20% for some of the parameter settings and up to 7% with the IDSA data-set. The inclusion of the 'do-not-always-update'(A) does not improve the results, in fact, it even decreases($p < .001$) the performance. This might be the effect of the not linearity of the trajectories; human behaviour is not regular especially in traffic congested cities. Hence detours from the ideal path a punished more, than moving into the right direction is rewarded.

The decrease in performance when using a path planner was not expected, since in real life people move using pedestrian path and the usage of a path planner seemed a good choice to model those trajectories. However, further analysis shows that the path planner incorporated u-turns rather than going backwards; hence the waypoints of POIs were always in front the tracked person.

It is clear that the number of POIs greatly influences the difficulty of the problem (Fig. 4b), but performance remains acceptable.

To investigate the robustness of the different rules regarding their parameter values, response screening is performed (Table 2). As expected, the angle α is the most influential parameter, but w_2 is also a significant factor. The remaining other variables only have a small impact on performance.

To further investigate robustness, the normalized performances of the best update rule (DE), and the most influential parameter (angle) are compared (Fig. 5). For both data-sets, the performance increases with the increase of α, indicating that the same parameter-settings can be used for both. The fact that the keep increasing is the effect of how the equations are implemented. α is linearly related with the increment of charge, therefore for big α the Δc is also

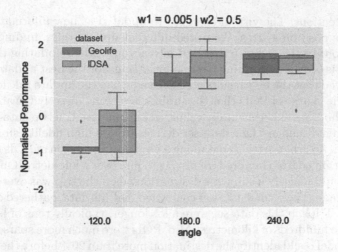

Fig. 5. Performances achieved by *DE* plotted using the best values for the most influencing factors. Every bar shows the median and the interquartile range of the performance over the median of 100 people tracked.

Table 2. Impact of the setting's factors to the final results.

Factor	Significance
α	$4.637 \cdot 10^{-26}$
s_1	0.359
s_2	0.220
w_1	0.005
w_2	$2.578 \cdot 10^{-15}$

bigger, and therefore heavily increases the potential, especially if the POI is in the same direction of the movement. This effect also caused the worsened performances for the short trajectories in the IDSA data-set.

5   Conclusion and Future Work

We presented a method to model the intended destination of a subject in real time, based on a trace of position information and prior knowledge of possible destinations (points of interest, or POIs). It does so without the need for prior analysis of habitual travel patterns, which puts it in contrast with existing work in this area. The knowledge of POI positions can, however, be derived from historical data, but it may also be available from other sources such as the organisation of one-off events or expert appraisal of the environment (e.g., tactical understanding of terrain in military contexts).

The method models the certainty of each POI by means of a virtual charge, resulting in an artificial potential field that reflects the current estimate of the subject's intentions. The virtual charges are updated as new information about the subject's position arrives. We tested different update rules, finding that it is important to take the distance to a POI into account when updating the charge. Close analysis of well-performing parameter settings for the best update rule considered showed that the best results are achieved when the update rule focusses on increasing the charge of POIs that the subject is moving directly towards.

The method was experimentally validated and analysed on two data-sets, both in an urban environment. One data-set derives from a high fidelity simulation of individuals in an urban setting containing a large concentration of POIs and where trip lengths ranged from hundreds of metres to just over a kilometre. On this data, the model could identify a subject's destination when the trip was over 90% complete on average. The second data-set concerned real-life data, gathered over a five-year period. Trips in this data-set were a lot longer, typically tens of kilometres, ranging up to hundreds of kilometres. Also, POIs were much more sparsely placed. Here, the model could identify the destination more than 20% before the end of the trip. Subsequent analysis bears out that the performance of our method seems to decrease linearly as the number of POIs increases.

While these results are encouraging, the model also has limitations. Mainly, the model struggles with short trips in an environment with many POIs and the accurate identification of a subject's destination is only possible after a substantial part of the trip has passed. Future work will investigate how to improve the update rules, for instance by using modified path planning algorithms to gauge a subject's distance to POIs. Also, prior knowledge of the relative likelihood of POIs and incorporating environmental information can likely improve the performance of the model. The model can further be extended to consider not only the final destination of each trip, but also waypoints on the subject's route. Additional research on different data-sets is needed to ascertain generalisation capability of this approach.

Overall, this paper offers a promising avenue of research towards methods to predict a subject's destination or even itinerary in real time without the need of analysing habitual movement patterns.

Acknowledgement. We thank SURFsara (www.surfsara.nl) for the support in using the Lisa Compute Cluster. The research for this paper was financially supported by the Netherlands Organisation for Applied Scientific Research (TNO).

References

1. Ashbrook, D., Starner, T.: Using GPS to learn significant locations and predict movement across multiple users. Personal Ubiquitous Comput. **7**(5), 275–286 (2003)
2. Nicholson, A.J., Noble, B.D.: BreadCrumbs: forecasting mobile connectivity. In: Proceedings of the 14th ACM International Conference on Mobile Computing and Networking, vol. 2, pp. 46–57 (2008)
3. Scellato, S., Musolesi, M., Mascolo, C., Latora, V., Campbell, A.T.: NextPlace: a spatio-temporal prediction framework for pervasive systems. In: Lyons, K., Hightower, J., Huang, E.M. (eds.) Pervasive 2011. LNCS, vol. 6696, pp. 152–169. Springer, Heidelberg (2011). doi:10.1007/978-3-642-21726-5_10

4. Do, T.M.T., Gatica-Perez, D.: Where and what: using smartphones to predict next locations and applications in daily life. Pervasive Mob. Comput. **12**, 79–91 (2014)
5. Sadilek, A., Krumm, J., Out, F.: Predicting long-term human mobility. In: 26th AAAI Conference on Artificial Intelligence, pp. 814–820 (2012)
6. Ziebart, B.D., Maas, A.L., Dey, A.K., Bagnell, J.A.: Navigate like a cabbie: probabilistic reasoning from observed context-aware behavior, pp. 322–331 (2008)
7. Liao, L., Patterson, D.J., Fox, D., Kautz, H.: Building personal maps from GPS data. In: International Joint Conference on Artificial Intelligence (IJCAI) Workshop on Modeling Others from Observations (2005)
8. Fallis, A.: Real-time travel path prediction using GPS-enabled mobile phones. J. Chem. Inf. Model. **53**(9), 1689–1699 (2013)
9. Lorenzo, G.D., Phithakkitnukoon, S., Horanont, T., Lorenzo, G.D., Map, A.-A.: Identifying human daily activity pattern using mobile phone data. In: Proceedings of the First International Conference on Human Behavior Understanding, pp. 14–25 (2010)
10. Ying, J.J.-C., Lee, W.-C., Weng, T.-C., Tseng, V.S.: Semantic trajectory mining for location prediction. In: Proceedings of the 19th ACM SIGSPATIAL International Conference on Advances in Geographic Information Systems - GIS 2011, p. 34 (2011)
11. Zheng, K., Zheng, Y., Yuan, N.J., Shang, S., Zhou, X.: Online discovery of gathering patterns over trajectories. IEEE Trans. Knowl. Data Eng. **26**(8), 1974–1988 (2014)
12. Hwang, Y.K., Ahuja, N.: A potential field approach to path planning. IEEE Trans. Robot. Autom. **8**(1), 23–32 (1992)
13. Howard, A., Mataric, M.J., Sukhatme, G.S.: Mobile sensor network deployment using potential fields: a distributed, scalable solution to the area coverage problem. In: Asama, H., Arai, T., Fukuda, T., Hasegawa, T. (eds.) Distributed Autonomous Robotic Systems, vol. 5, pp. 299–308. Springer, Tokyo (2002)
14. Parunak, H., Purcell, L., Six, F., Station, N., O'Connell, M.: Digital pheromones for autonomous coordination of swarming UAV's. In: 1st UAV Conference, Infotech@Aerospace Conferences (2002)
15. Mottaghi, R., Vaughan, R.: An integrated particle filter and potential field method applied to cooperative multi-robot target tracking. Autonom. Robots **23**(1), 19–35 (2007)
16. Helble, H., Cameron, S.: 3-D path planning and target trajectory prediction for the Oxford aerial tracking system. In: Proceedings of the IEEE International Conference on Robotics and Automation, pp. 1042–1048 (2007)
17. de Jong, S.; Klein, A., Smelik, R., van Wermeskerken, F.: Integrating run-time incidents in a large-scale simulated urban environment. In: Proceedings of the 2016 International Conference on Autonomous Agents & Multiagent Systems, pp. 1401–1402 (2016)
18. Zheng, Y., Li, Q., Chen, Y., Xie, X., Ma, W.-Y.: Understanding mobility based on GPS data. In: Proceedings of the 10th International Conference on Ubiquitous Computing - UbiComp 2008, vol. 49, p. 312 (2008)
19. Zheng, Y., Zhang, L., Xie, X., Ma, W.-Y.: Mining interesting locations and travel sequences from GPS trajectories. In: Proceedings of the 18th International Conference on World Wide Web - WWW 2009, vol. 49, p. 791 (2009)
20. Zheng, Y., Xie, X., Ma, W.: GeoLife: a collaborative social networking service among user, location and trajectory. IEEE Data Eng. Bull. **33**(2), 32–40 (2010)

Robust Bi-objective Shortest Path Problem in Real Road Networks

Christian Cintrano$^{(\boxtimes)}$, Francisco Chicano, and Enrique Alba

Departamento de Lenguajes y Ciencias de la Computación,
Universidad de Málaga, Andalucía Tech., Málaga, Spain
{cintrano,chicano,eat}@lcc.uma.es

Abstract. Road journeys are one of our most frequent daily tasks. Despite we need them, these trips have some associated costs: time, money, pollution, etc. One of the usual ways of modeling the road network is as a graph. The shortest path problem consists in finding the path in a graph that minimizes a certain cost function. However, in real world applications, more than one objective must be optimized simultaneously (e.g. time and pollution) and the data used in the optimization is not precise: it contains errors. In this paper we propose a new mathematical model for the robust bi-objective shortest path problem. In addition, some empirical studies are included to illustrate the utility of our formulation.

Keywords: Robustness · Traffic road network · Bi-objective shortest path · Multi-objective optimization

1 Introduction

Road trips are an inherent part of our modern lifestyle. To give a quick example, only in U.S.A., citizens spend almost two hours a day driving. We must add to this time the economic (fuel, vehicle maintenance, ...) and psychological (stress, anxiety, ...) costs. For this reason, it is not surprising that the problem of finding the shortest or fastest paths between two points in the city is a popular and well studied problem in the scientific literature.

A rather generalized way of modeling the city's road network is to treat it as a weighted directed graph. The streets are the arcs of this graph, the city intersections are the nodes, and the weights could be the length of the road, speed limits, etc. If we use this model, well-known algorithms like Dijkstra [1] or A* [2] are useful even for commercial applications. However, these models have a very simplistic view of the city, offering as a solution a single route that (generally) only minimizes time (or distance).

But time is not the only objective for the citizens. Fuel, modern environmental concerns, or reduced traffic jams are also new goals to be taken into account for smart mobility in cities. A good point to start with new formulations is to consider the bi-objective shortest path (BSP) problem, that has received special attention in recent years both in academy and industry [3].

© Springer International Publishing AG 2017
E. Alba et al. (Eds.): Smart-CT 2017, LNCS 10268, pp. 128–136, 2017.
DOI: 10.1007/978-3-319-59513-9_13

To make a difference, it is important that the proposed solutions for the BSP problem are applicable in the real world. This requires taking into account real-time and precise information on the state of the roads, traffic jams, flows of vehicles, etc. The challenge of the use of these data sources falls in its dynamic nature, e.g., traffic in rush hour differs from that in the evening, or in its stochasticity, e.g., an accident happens on the road. These are the reasons why a useful solution (algorithm) must take these imprecisions into account and avoid providing routes that, for example, can cause by themselves additional delays in the arrival to the destination. The ability of a solver to tolerate these inaccuracies is called *robustness*. And a variant of the BSP problem that takes into account inaccuracies in its input data will be called *robust bi-objective shortest path problem* (RBSP problem).

In this article we will present a new model of RBSP problem that deals with the robustness from a stochastic point of view. We assume that the input data behaves as probability distributions. Also, we decided work with the bi-objective version of SP problem given that most of the existing instances for this kind of problem optimize only one or two objectives. Therefore BSP problem is the natural choice on which to make our robust model.

The article is structured as follows: Sect. 2 shows the basic formulation of the RBSP problem. Section 3 describes our proposal for modeling the robustness of the problem. Section 4 presents some empirical results of our selected model. Section 5 presents alternative formulations of the robustness for the RBSP problem existent in the state-of-the-art. Finally, Sect. 6 presents the main conclusions of this article and the lines of future work.

2   Basic Formulation

Before presenting our model, it is necessary to give some basic notions about the BSP problem and about the robustness in optimization problems. Then, we formulate the mono-objective and bi-objective shortest path problem. Later, general considerations about robustness in optimization problems are presented.

2.1   Mono- and Bi-objective Shortest Path Problems

Let $G(N, A)$ be a directed graph, where N is the set of nodes, and A the set of arcs between nodes, $A \subseteq N \times N$; we define a path $p = p_1, p_2, \ldots, p_k$ as a list of nodes that: $\forall p_i, p_{i+1} \in p, 0 \leq i \leq k - 1, (p_i, p_{i+1}) \in A$. We define $\mathcal{P}_{s,e}$ as the set of all possible paths between a start node s and an end node e.

We can define a cost function $C : A \to \mathbb{R}^+$ in the graph G. This function assigns a non-negative numeric weight value to each edge of the graph. In order to simplify the formulation, we write $C((i, j)) = c_{ij}$ as the cost of the arc (i, j).

The path between two given nodes s and e in the graph that minimizes the total cost of the path is the result of solving the following problem:

$$\min_{p \in \mathcal{P}_{s,e}} \boldsymbol{z}(p) = \sum_{(i,j) \in p} c_{ij} \tag{1}$$

where $z(p)$ is the objective function, s and e are (respectively) the start and end nodes of the path. Abusing of notation we write $(i, j) \in p$ when the arc (i, j) appears in the path p.

We will consider, without loss of generality, that the objective (components of the vector function) are to be minimized. Next, we include the definition of some standard multi-objective concepts to make the paper self-contained.

Definition 1 (Dominance). *Given a vector function* $\mathbf{f} : \mathcal{P}_{s,e} \rightarrow \mathbb{R}^d$*, we say that solution* $x \in \mathcal{P}_{s,e}$ *dominates solution* $y \in \mathcal{P}_{s,e}$*, denoted with* $x \prec_{\mathbf{f}} y$*, if and only if* $f_i(x) \leq f_i(y)$ *for all* $1 \leq i \leq d$ *and there exists* $j \in \{1, 2, \ldots, d\}$ *such that* $f_j(x) < f_j(y)$*. When the vector function is clear from the context, we will use* $\prec$ *instead of* $\prec_{\mathbf{f}}$*.*

Definition 2 (Pareto Optimal Set and Pareto Front). *Given a vector function* $\mathbf{f} : \mathcal{P}_{s,e} \rightarrow \mathbb{R}^d$*, the* Pareto Optimal Set *is the set of solutions* P *that are not dominated by any other solution in* $\mathcal{P}_{s,e}$*. That is:*

$$P = \{x \in \mathcal{P}_{s,e} | \nexists y \in \mathcal{P}_{s,e}, y \prec x\}. \tag{2}$$

The Pareto Front *is the image by* $\mathbf{f}$ *of the Pareto Optimal Set:* $PF = \mathbf{f}(P)$.

Definition 3 (Set of Non-dominated Solutions). *Given a vector function* $\mathbf{f} : \mathcal{P}_{s,e} \rightarrow \mathbb{R}^d$*, we say that a set* $X \subseteq \mathcal{P}_{s,e}$ *is a set of non-dominated solutions when there is no pair of solutions* $x, y \in X$ *where* $y \prec x$*, that is,* $\forall x \in X, \nexists y \in X, y \prec x$.

Now we can define the bi-objective shortest path (BSP) problem. This problem was defined by Hansen [4] in 1980. Here is a simple definition of the BSP problem:

Definition 4 (Bi-objective Shortest Path). *Let,* $C = (C^1, C^2)$ *a pair of cost functions, and* $c_{ij,k}$ *the cost of the arc* (i, j) *in objective* k*. The bi-objective shortest path problem are defined as follows:*

$$\min_{p \in \mathcal{P}_{s,e}} \mathbf{z}(p) = \sum_{(i,j) \in p} (c_{ij,1}, c_{ij,2}) \tag{3}$$

2.2 Robustness

In real problems, one of the biggest problems is the many inaccuracies in the data. Changes in the environment, noisy sensors and actuators, and lack of knowledge of the problem are some examples of the issues that must be faced by algorithms and applications that want to be useful in the real world. These events led to a new optimization paradigm called robust optimization [5]. The lack of precision can be modeled in two different ways, as explained in [6]: (i) stochastic, which models inaccuracies assuming a random variable for each variable; (ii) robust, assume that the value of each variable is obtained from an uncertain set of values.

In this paper we will use probability distributions instead of fixed values in the cost function of the graph. In this work, we call *robustness* to the ability of dealing with inaccuracies.

3   Robustness Model

There are multiple ways to apply robustness to the BSP problem. However, they all have a number of limitations, either by simplifying the real world, or by generating complex objective functions. We here propose a simple model of robustness for this problem based on treating the costs as probability distributions. With this approach, we want to generate a simple objective function that is able to deal with not exact data input.

In real world, much data is uncertain. In the routing of vehicles, the speed of vehicles, the time, the amount of gases emitted, etc. depend on multiple factors. Some examples of such factors are: the hour of the day, the degree of congestion of the road/lanes, the model of vehicle used, or the driver's profile in terms of acceleration up/down. For this reason, for the same trip, these environment variables can take multiple values.

Despite this apparent randomness, environment variables often have a temporary and stationary behavior. Rush hour traffic is usually similar in the same period of the year. Given this behavior we could interpret, for example, travel time as a probability distribution with μ, the average travel time at any hour of the day, and σ^2, its variance. We assume that random variables are independent.

For our RBSP problem, we will transform our fixed arc costs $c_{ij,k}$ into random variables $\hat{c}_{ij,k}$. This transformation of the kth-cost function into the new function will depend on the knowledge of the particular real world variable. In a similar way, we change our objective function $z(p)$ in a new $\hat{z}(p)$

$$\hat{z}(p) = \sum_{(i,j)\in p} \hat{c}_{ij,k} \tag{4}$$

A random variable can have many values in theory. However, with regards to road travel, many factors tend to have similar values. A typical way to be more flexible in the data is consider, e.g., travel time, within a range. But, unforeseen events can always occur that cause the data come out of that range. One way to take all this information into account is to use probability distributions that most closely resemble real behavior. We will characterize each probability distribution by two functions: $\mu(\hat{c}_{ij,k})$ that return the average of the probability distribution, and $\sigma^2(\hat{c}_{ij,k})$ that returns the variance as output. Abusing of notation we write $\mu(\hat{c}_{ij,k})$ as $\mu_{ij,k}$, that means the average of the random distribution of the arc (i,j) in the objective k. In a similar way we write $\sigma^2(\hat{c}_{ij,k})$ as $\sigma^2_{ij,k}$.

For an ideal road trip, its associated cost must be minimal and have the least variation possible. Based on this idea we use, in Eq. 5, our robust objective function to define the RBSP problem.

$$\min_{p\in\mathcal{P}_{s,e}} z^R(p) = \sum_{(i,j)\in p} (\mu_{ij,1}, \mu_{ijm,2}, \sigma^2_{ij,1}, \sigma^2_{ij,2}) \tag{5}$$

The values of μ and σ^2 are the new costs associated to each arc of the graph. In this way, we transform our bi-objective problem into another one with four objectives, being able to apply any general algorithm that solve problems with four objectives.

4   Experimental Study

In this section we will present some experiments to illustrate the use of our model. We will use the real map of the city of Málaga (Spain) to show how our approach is applicable to the real world.

4.1   Algorithms

We test the impact of applying robustness to our problem by using two types of algorithms. On the one hand, we use the mono-objective algorithms Dijkstra and A*, because they are typically used in the state-of-the-art and applications. On the other hand, we use an algorithm to obtain the Pareto optimal set in our four-objective robust model. The selected the state-of-the-art multi-objective algorithm is PULSE [7]. This is an exact recursive algorithm based in aggressive pruning strategies. We implemented PULSE in the Java programming language and included the modifications that the authors proposed in the article to extend it to four objectives.

4.2   Málaga City (Spain)

The city of Málaga is one of the Spanish cities pioneering in smart cities initiatives. It is an example of medium-size European city. Figure 1 shows a map of Málaga. The graph used in this study to model the city was obtained from Open Street Map[1]. This graph had a preprocessing step to became strongly connected and the resulting graph has 45,410 nodes and 118,388 arcs. Thanks to this, our results would be compatible with applications that make use of real maps of Open Street Maps. In this work, since we only want to validate the proposed model, the weights of the arcs were randomly generated.

Fig. 1. Map view Málaga (Spain) from Open Street Map

---

[1] Official web site: https://www.openstreetmap.org.

4.3 Methodology

In this section we describe the methodology follows in our empirical study. To illustrate the behavior of our model we are going to use algorithms that return solutions without regard to robustness (Dijkstra and A*) and others whose outputs take into account the robustness (PULSE). Because, Dijkstra and A* are mono-objective algorithms, we use weighted sums to take into account two objectives. In both cases, we only assume fixed data in these two mono-objective algorithms, so we use the weighing: $\alpha_1^R + (1 - \alpha)z_2^R$ with the weight α getting the different values: $\alpha = 1$, $\alpha = 0$, $\alpha = 0.5$. With this weighting we want to get a variety of solutions.

We run the algorithms which inputs are a graph (Málaga) and a start and end nodes. In the graph of Málaga we selected a total of five pairs of random nodes. Each node is numbered between 0 and 9. For each pair (i, j) we calculate two paths: one from i to j, and another from j to i.

These paths generate a total of ten instances on which we run the algorithms. We will measure the execution times of each algorithm, because it is desirable that the algorithm be fast, and the four objectives (z^R) of each solution obtained: the Pareto set in the PULSE algorithm and the single solution in the mono-objective algorithms. Get z^R will allow us to compare if one solution is more robust than another. In this paper, we will assume that one solution (path) p^1 is more robust than another p^2 if $z_3^R(p^1) < z_3^R(p^2), z_4^R(p^1) < z_4^R(p^2)$.

4.4 Empirical Results

Since the representation of four objective variables is a complex problem, we have opted for a representation in the form of value ranges as it is showed in the Fig. 2. In this, an example of the Pareto front of one of our test paths (3,2) is displayed. The boxes are centered on z_1^R (x-axis) and z_2^R (y-axis) values; and its width of each box expresses the z_3^R (the y-axis is analogous to the x-axis). We can see that Dijkstra is not within the front obtained by PULSE. A*, on the other hand, is on the front, but it is only one of the 21 solutions in the front.

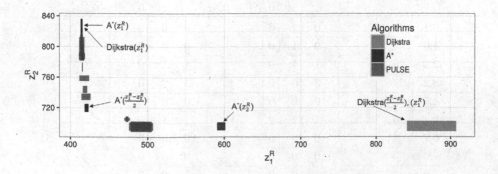

Fig. 2. Pareto front for one path of the Málaga City

Table 1. Computational time in seconds of each algorithm

Path	Dijkstra			A*			PULSE
	z_1^R	z_2^R	$0.5z_1^R + 0.5z_2^R$	z_1^R	z_2^R	$0.5z_1^R + 0.5z_2^R$	
0,1	0.53	4.45	0.14	0.50	4.02	0.29	45.65
1,0	0.05	2.66	0.06	0.07	5.72	0.10 ·	82.89
2,3	1.23	4.29	0.72	1.48	4.09	1.81	4879.25
3,2	0.31	1.26	0.48	0.74	3.39	0.98	3641.96
4,5	1.05	2.87	1.11	0.92	1.58	1.15	1696.29
5,4	0.96	1.01	0.75	1.04	2.63	1.72	11950.93
6,7	1.14	1.26	1.73	1.25	0.81	1.51	1192.76
7,6	0.24	0.14	0.13	0.17	0.33	0.26	1616.03
8,9	0.49	4.12	0.43	0.26	9.87	0.88	115.98
9,8	0.04	0.64	0.04	0.06	2.22	0.28	171.19

Next we will analyze some parameters measured in the different paths. Table 1 shows the execution times of each algorithm to compute each path. As we would expect, PULSE is slower than the other two algorithms. The differences between execution times in the PULSE algorithm are because the proximity between the start and end points. However, while the mono-objective algorithms only found one solution, PULSE finds the whole Pareto set of solutions.

In order to compare the robustness between the obtained solutions, we calculated the number of solutions in the Pareto set whose objectives z_3^R and z_4^R (variances of the two probability distributions) are simultaneously strictly smaller than the solutions obtained by the three versions of Dijkstra and A*. In the Table 2 we observe that almost three solutions with less variability are obtained.

Table 2. Pareto size and number of solution obtained by PULSE more robust than mono-objective algorithms

Path	Pareto set size	Dijkstra			A*		
		z_1^R	z_2^R	$0.5z_1^R + 0.5z_2^R$	z_1^R	z_2^R	$0.5z_1^R + 0.5z_2^R$
0,1	4	0	1	0	0	0	0
1,0	4	0	1	0	0	0	0
2,3	26	1	1	0	1	1	0
3,2	21	3	0	0	3	2	0
4,5	5	0	0	0	0	0	0
5,4	6	1	1	0	1	1	0
6,7	4	0	0	0	0	0	0
7,6	3	0	0	0	0	0	0
8,9	6	1	0	0	1	0	0
9,8	24	8	0	0	8	0	0

While the computing time is higher, there are situations where it is preferable to obtain more predictable routes, i.e., travel to airport or job interviews, salesman that should visit various commercial establishment, etc.

5   Related Work

In tis section we present some works related to the application of robustness to SP and BSP problem. First, the problem of the mono-objective SPP with robustness is well studied in the literature. There are many models that add robustness to this problem in different ways.

In [8] a random factor is added as a delay. This adds some imprecision in the weight of the bow. They also describe how to use the knowledge of the probability distribution of this random variable. The authors solved this problem by applying a strategy similar to the minmax regret, using the supreme of the probability distribution of the delay.

In [9], the authors modeled robustness using confidence intervals. Although they tackled the mono-objective problem, they transformed it into a multi-objective problem by moving from exact values to confidence intervals.

Other authors have started off from the existence of a finite set of possible combinations of values that can take the weights of the arcs. These sets are called scenarios. An example of this type of robustness for the SP problem is [10]. In this paper, the authors used a minmax regret strategy to find a solution that minimizes cost in the worst case scenario.

There are many papers that analyze robustness in bi-objective problems from a theoretical point of view [11,12]. However, in all these articles the robustness is only applied to one single objective. The mean strategy for finding a solution is the minmax regret as in the mono-objective version of the problem.

6   Conclusions

In this paper a new model for the robust bi-objective shortest path problem has been presented. Our purely multi-objective approach is distinct from the rest of commonly used techniques. Thanks to our approach we can get a set of solutions with different degrees of robustness and quality with respect to two metrics. In addition, we have empirically illustrated our proposal comparing the results with those typical algorithms of the state-of-the-art and show how we can get more robust solutions than them.

As a future work, we will use real data instead of random weights and different maps for solve the RBSP problem and so analyze the behavior of our model over different set of data. Besides, we will compare our approach with another models and implemented different algorithms to resolve the BSP and RBSP problem. Also, we will apply this methodology of dealing with the robustness to other problems in smart cities. We will also try not only to treat robustness, but also to quantify its degree of use in different algorithms and applications.

Acknowledgements. This research was partially funded by the University of Málaga, Andalucía Tech, and the Spanish Ministry of Economy and Competitiveness and FEDER (grant TIN2014-57341-R).

References

1. Dijkstra, E.W.: A note on two problems in connexion with graphs. Numer. Math. **1**(1), 269–271 (1959)
2. Hart, P.E., Nilsson, N.J., Raphael, B.: A formal basis for the heuristic determination of minimum cost paths. IEEE Trans. Syst. Sci. Cybern. **4**(2), 100–107 (1968)
3. Cintrano, C., Stolfi, D.H., Toutouh, J., Chicano, F., Alba, E.: CTPATH: a real world system to enable green transportation by optimizing environmentaly friendly routing paths. In: Alba, E., Chicano, F., Luque, G. (eds.) Smart-CT 2016. LNCS, vol. 9704, pp. 63–75. Springer, Cham (2016). doi:10.1007/978-3-319-39595-1_7
4. Hansen, P.: Bicriterion path problems. In: Fandel, G., Gal, T. (eds.) Multiple Criteria Decision Making Theory and Application. LNE, vol. 177, pp. 109–127. Springer, Heidelberg (1980)
5. Ben-Tal, A., El Ghaoui, L., Nemirovski, A.: Robust Optimization. Princeton University Press, Princeton (2009)
6. Ide, J., Schöbel, A.: Robustness for uncertain multi-objective optimization: a survey and analysis of different concepts. OR Spectrum **38**(1), 235–271 (2016)
7. Duque, D., Lozano, L., Medaglia, A.L.: An exact method for the biobjective shortest path problem for large-scale road networks. Eur. J. Oper. Res. **242**(3), 788–797 (2015)
8. Cheng, J., Lisser, A., Letournel, M.: Distributionally robust stochastic shortest path problem. Electr. Notes Discrete Math. **36**, 511–518 (2010)
9. Hasuike, T.: Robust shortest path problem based on a confidence interval in fuzzy bicriteria decision making. Inf. Sci. **221**, 520–533 (2013)
10. Pascoal, M.M., Resende, M.: The minmax regret robust shortest path problem in a finite multi-scenario model. Appl. Math. Comput. **241**, 88–111 (2014)
11. Ehrgott, M., Ide, J., Schöbel, A.: Minmax robustness for multi-objective optimization problems. Eur. J. Oper. Res. (1), 17–31
12. Kuhn, K., Raith, A., Schmidt, M., Schöbel, A.: Bi-objective robust optimisation. Eur. J. Oper. Res. **252**(2), 418–431 (2016)

Smart Urban Mobility from Expert Stakeholders' Narratives

Daniel Lopatnikov[✉]

Department of Sociology, Public University of Navarra, Pamplona, Spain
daniel.lopatnikov@unavarra.es

Abstract. First World developed regions are usually those which pioneer technological and social advancements that later spread over the rest of the globe. Mobility has a considerable impact on cities and the regions which contain them. "Smart cities" are expected to promote denser city centres where not only private car needs are taken into account. Decision-making is crucial for boosting social cohesion and inclusiveness through the right social appropriation of technological progress. The Spanish region of Navarra and its capital city Pamplona are currently experiencing profound transformations which aim at integrating "smart city"-related innovations. In spite of the central relevance of the automobile industry for its economy, Pamplona is making efforts for consolidating alternatives to its private car-centred mobility. Intermodal transportation options offer a more sustainable and socially inclusive mobility-paradigm. This paper incorporates an analysis of the narratives and discourses of some of the key stakeholders in play, such as politicians, transportation experts and engineers. Their main viewpoints are examined from a sociological approach as they could predict new collective trends valid not only for this mid-size city but also for many analogous cases.

The twentieth century idealisation of suburban citizens who live in residential areas next to a city and drive everywhere through wide roads in their private automobiles seems to be fading. The electrification of both private vehicles and public transportation is also regarded as imperative in tackling city pollution. Social and academic debate on how to successfully manage this paradigm shift is already being raised.

Keywords: Mobility-Paradigm shift · Urban-transport agenda · Socio-technical transition

1 Introduction

This paper is based on case study research obtained over six months of fieldwork. The experts' narratives analysed here have been gathered in Pamplona (Spain) but could also be valid for many other comparable medium-size cities in the First World which have privileged access to technology. This type of cities are vital to any developed society's economy and overall structure (for mid-size European cities such as Pamplona, Oviedo, Montpellier, Salzburg, Brugge, Eindhoven... see Giffinger in [1]).

© Springer International Publishing AG 2017
E. Alba et al. (Eds.): Smart-CT 2017, LNCS 10268, pp. 137–143, 2017.
DOI: 10.1007/978-3-319-59513-9_14

The first part of this paper will be dedicated to depicting a brief outline of the impact which the latest technologies are about to bring to modern First World cities. This phenomenon, called by some authors "the Fourth Industrial Revolution" (see Schwab in [2]), is expected to raise new challenges in terms of managing the intrinsic complex relationship between state of the art technical engineering and social engineering, meaning that both exact sciences and social sciences need to be put to work together in order to avoid an unequal distribution denying the benefits of technical progress to those citizens unable to make the most of them.

Secondly, the discussion will point out the main characteristics of the mobility transition which cities located in the developed world are experiencing nowadays. This is closely associated with the introduction of new trends brought from different engineering fields and is partly responsible for giving shape to what we tend to call "smart cities", even though this term remains to be fully defined due to its newness. Most modern cities have undergone a progressive transformation which first led to the boom and later the decay of both the collective (based on mass means of transport such as the bus or the tram) and the motored-privatised (based on a car per household, and afterwards on a car per adult person) paradigms of mobility. Intermodal transportation systems seem to be emerging as the in-vogue and most sustainable alternative for city mobility.

The central part of the discussion will focus on the demands faced by a medium-size European regional capital, a case which could be quite similar to any comparable-size European or, to a lesser extent, world city if they are not too distant in terms of economic advancement and overall quality of life. It is of key relevance to examine the main stakeholders' viewpoints as not only the regular citizens' approaches to "smart city"-related issues define the unfolding of a "smart city". The assessments and decision-making of these experts will co-design the shape of future city mobility.

Finally, a set of conclusions based on the discussion will be drawn up with the intention of highlighting the basic contributions made and the pertinence of putting all significant stakeholders to work together with regular citizens to collectively address this important issue which is not only a technical matter but also a sociological one.

2 Case Study and Experts' Profiles

Pamplona is a city with a quarter of a million inhabitants located in the north of Spain, less than a hundred kilometres from the French border. It is the economic and political capital of Navarra, a province which is situated in one of the most prosperous parts of Spain. Navarra and Pamplona score over the Spanish averages in all HDI (Human Development Index)-related and income rankings and are usually close to Northern European averages. If not tackled for sustainability purposes, *"economic development… results in higher incomes and improved road transport infrastructure, which may encourage a shift to private transportation and longer average vehicle driving distance"* [3]. In fact, the suburban perimeter of Pamplona has notably spread over recent decades and distances have risen. Urban field experts Camarero and Oliva [4] conclude about this city that *"in view of its medium size, and its dispersion based almost exclusively on increasing private automobile use, it offers a paradigmatic case for study."*

It is also relevant to remark that the automobile industry is financially crucial for Pamplona, Navarra and Spain. Spain ranks in the top ten of the largest automobile producers in the world and its vehicle exports account for around twenty percent of the value of its exports. The weight of the car industry is even greater for Navarra and Pamplona, as one of the most thriving factories is located in this regional capital. It is estimated that the industrial sector is worth more than a quarter of the regional GDP and most of it is closely linked (through the factory itself and the many component and service providers) to the assemblage of the Volkswagen Polo in Pamplona.

However, even though the status of cars in "smart cities" remains controversial, the city of Pamplona was ranked in a research carried out by Siemens in 2012 as the 5th most sustainable city in Spain. This evaluation took into account aspects such as mobility, CO2 emissions and air purity. In the same year, the consulting agency IDC determined that Pamplona was in the top ten of the Spanish Smart Cities. The most interesting thing is that this was known even before the city itself made public its specific agenda for progressively becoming "smart". The council recently published a so called Smart City Pamplona strategic plan defining its goals for the coming years.

In accordance with this given context, a number of experts representing some of the key stakeholders in play have been selected for the case study fieldwork. During the last half year several in-depth interviews with experts have been conducted to analyse their main thoughts and narratives. This was intended to enrich a research project which would not only consist of data, academic literature and official policy statements. The in-depth interviews did not keep to a strict common script as it was more viable to adapt the questions to each of the participants' areas of expertise.

The final selection of in-depth interviewees for illustrating this paper consists of two members of the Public University of Navarra's Institute of Smart Cities, as well as two relevant experts from the local transportation administration and the city mobility councillor. The first two are F.J. Falcone and J. Faulín, who are university professors and experts in electrical engineering and statistical investigations, respectively. The second two are R. Bujanda, head of the Department for the Modernisation of the Transportation Services in Navarra, and P.L. Rezusta, transportation planning expert at the local taxi office. Finally, A. Cuenca is Pamplona's current mobility councillor and member of a local left-wing-ideology party called Aranzadi. Several other stakeholder institutions have provided significant extra information for further analysis of the points made by these interviewees and new key players such as the city mayor have agreed to collaborate with their participation in the following interviews.

3   Main Discussion

"Smart cities", new ways of urban transportation and mobility and engineering progress are very tightly linked. All sociological approaches to this matter must take into account the degree of introduction of the latest technological advancements, as those shape the circumstances and potential mobility of all civil-society actors. Schwab argues that we are currently experiencing changes which are *"… historic in terms of their size, speed and scope"* [2]. The Internet of things, Big Data accumulation and management, online platforms for the sharing economy, electric and online connected driverless vehicles… These radical changes are new challenges for developed cities in a globalised society. The complex nature and interconnectedness of these transformations, whether called the "Fourth Industrial Revolution" or not, make it *"…critical that we invest attention and energy in multistakeholder cooperation across academic, social, political, national and industry boundaries"* [2]. It is crucial that average citizens and not only the richest ones get access to the benefits brought about by the new technologies as the overall health and cohesion of a city will importantly depend on this social appropriation.

In terms of mobility, this evolution is giving shape to a profound transition from older paradigms to the newest trends. After the collective transportation model decay (buses, trams, trains), the privately-owned car arose as the mainstream form of city and general mobility. For Urry, the car became *"… the quintessential manufactured object produced by the leading industrial sectors and the iconic firms within twentieth-century capitalism (Ford, General Motors, Volvo, Rolls-Royce, Mercedes, Toyota…"* [5]. Not only did the car succeed as an alternative for going from one place to another, it also pushed modern cities to modify their planning and structure: *"Indeed the automobile and the suburban family home, are arguably the primary commodities that fuelled mass consumption throughout the twentieth century"* [6]. This well-known model which involves both a massification of private transport and an urban sprawl of the city appears to be fading. Tailpipe emissions, the invasion of space, noise pollution, the high cost of electrification… Cities are becoming more aware than ever of the dangers of perpetuating a private car-centred paradigm of mobility.

Intermodal alternatives for transportation are thought to be the answer to current private car-related problems. This does not mean abandoning the idea of owning a car. Instead, intermodality stands for the optimal combination of several means of transport so that the still popular "a car per person for going everywhere" mindset is left aside due to its social drawbacks. Many modern cities are increasingly opting for denser walkable city centres, bicycle and bus lane infrastructures and other measures to discourage heavy car use. The former major of Bogotá became one of this paradigm's iconic supporting leaders when he claimed that a developed society is not one where the poor have cars but one where the rich use public transport.

The expansion of the Internet of things is one of the first matters that experts outlined as relevant. Smart cities and smart mobility mean constant connectedness. Engineering professor F.J. Falcone suggests that ITS (Intelligent Transport Systems) will very soon be able to improve the safety of our trips. He offers an example: *"If you had an accident, your car itself will spread it online so that other drivers and the authorities instantly have that information. You will also receive constantly updated weather related route tips."* This will all depend upon a permanent connection between vehicles and city infrastructure and will not only benefit cars but also public transport vehicles.

It is expected that all city vehicles will progressively become electric as well as being connected online. This seems to be a sequential pattern for which internal combustion engines are losing prevalence against hybrid ones and the latter will then be substituted for pure electric vehicles. In hybrid vehicles, *"the smaller ICE reduces fuel consumption, and operating the engine in its optimal range greatly improves tailpipe emissions"* [3]. Full electrification is expensive in terms of infrastructure and current vehicle models do not offer convenient autonomy yet. It is not even clear how to deal with the issue of electricity production, as renewable energy supply is not always guaranteed.

In any case, experts agreed that mobility electrification is reaching modern cities and Pamplona is not an exception to this global trend. F.J. Falcone estimates that in a 5 or more likely 10 year time period, greater battery capability and lower car prices will lead to a massive popularisation of electric vehicles inside cities. Even Tesla is already making their models more affordable, even though they remain too expensive for the average European citizen. Stakeholders consider that a boom of electrified mobility will need proper legislation (tax incentives, priority during pollution peaks, free parking, etc.) and new infrastructure networks. F.J. Falcone blames this latter point for the lack of progress, arguing that we will not find a plug-in point in each parking place in Pamplona and our communities will not be willing to pay our own electricity bills if we want to buy an electric car. Local taxi service manager P.L. Rezusta finds it logical that considering these circumstances only one taxi is currently fully electric in Pamplona. Hybrids are a lot more popular for the moment, having reached a considerable figure of 84 vehicles (there still being 229 diesel-engine taxis on the roads). Rezusta predicts a short-term increase in the number of hybrid vehicles and a long-term emergence of fully electric ones in accordance with the region's target for sustainability.

But not everyone is for the electrification of city transportation. Mobility councillor A. Cuenca claims that, for him, *"only electric bicycle mobility is a clearly successful advancement as it can help older people cope with hill-climbing. Instead, electric cars would only solve the city's pollution problems as they would keep invading space and would not guarantee pollution-free energy creation unless laws enforced it."* Besides, some find it dangerous to have nearly noise-free vehicles on the streets as pedestrians could easily be run over. Institute of Smart Cities member J. Faulín comments that people might need to be re-educated so that a lack of noise in the roads benefits city inhabitants and is not associated with a total absence of vehicles with its attendant risks.

The progressive electrification of transport (concerning mainly private vehicles, as fully electric public transport is a lot less controversial) is not seen as an ideal option by a few reluctant stakeholders. Some narratives are far from agreeing with the mainstream positions and stress the disadvantages of perpetuating a car-centred paradigm, even though the car could be free of pollutants. Councillor Cuenca suggests that positions are closely related to political ideologies: Right-wing parties are more prone to defend a car-centred society in which cars symbolise the individual's freedom of movement. He even blames Navarra's regional left-wing government for promoting fully-electric vehicles and battery charging stations. He claims to be more representative of the real left-wing positions when he opts for better bike and bus lanes and wants to limit the amount of cars to a few lanes in Pamplona's principal avenues. His ideas are already stirring up a heated debate among citizens of Pamplona, politicians and the local media.

Public transport faces great difficulties when efforts are made to promote its popularity. The key problem is that it is very hard to compete with private cars in terms of general convenience, and most citizens in Pamplona can afford to own one. Experts definitively conclude that public transport users are not representative of the city's population. This is because some social minorities are overrepresented. Regional specialist R. Bujanda reckons that public transport is predominantly used by women, the young and people with a physical impairment to drive. Cuenca remarks that other disadvantaged citizens are also likely to be seen on Pamplona's bus network: Immigrants and the uneducated. Even though he himself has raised concerns over this situation and has proposed replanning some city centre avenues to create a Bus Rapid Transit (BRT)-like network, the city's public transit system remains far from seducing the majority of the people.

When requested to opt for a better mobility paradigm than the problematic private car-centred one, most stakeholders deem intermodal transportation to be the ideal solution. A flawless city mobility network would mean a balanced combination of all options of transportation available: a restricted use of the private car just for specific activities such as weekend trips or shopping, cycling to nearby places, convenient public transport for longer urban distances, car-sharing, etc. Professor F.J. Falcone offers a brief example: *"Intermodality means unifying the whole network of transportation. You should be able to land near Barcelona, take a bus to Barcelona and then a train to Pamplona. Here, you could rent a bike to get to your nearby final destination. All this could be paid through a single mobile app."* Even though these kinds of paying platforms already exist, he admits that it is still too difficult for public and private stakeholders to success-fully cooperate. Small steps are being taken in this direction: R. Bujanda claims that *"new mobility cards valid both for inner-city and regional buses are just about to be released. Pamplona and Navarra will be totally unified in terms of public transit."*

If the different elements of the previous discussion were considered as a whole, a shift of paradigm in Pamplona's mobility patterns could easily be identified. A final remark was made by smart city researcher J. Faulín, who suggested that all these changes would not only mean sustainability but would lead to palpable gains in mental health: *"We are slaves of our cars. They need maintenance, insurance, fuel... In the near future we must be just users who want to get from one place to another in the fastest and best possible way. We will be considerably freer and less worried about ownership."* City mobility would be less car-reliant, flexible and more affordable for all as public transport in Pamplona is subsidized and unprofitable (as showed by R. Bujanda). In spite of its cost, an improvement of any city's transport network would be beneficial, as *"it affects both the quality of life of individuals and the equity and cohesion of a society as a whole"* [7].

4   Conclusions

This paper has highlighted the relevance of having access to key stakeholders' visions in order to being able to anticipate when and how a city and its people will incorporate further "smart city"-related progress and technological innovations. The central role of urban mobility has been the focus of attention here, as it importantly shapes not only transportation systems but also infrastructures, urban planning and lifestyle choices.

Some narratives and discourse analysis from a very representative case study have been included. Pamplona is paradigmatic of the current city-mobility transition as its main actors are trying to manage the challenges faced by a region which is disproportionately private car-dependent and which is at the same time a place of vital importance for the automobile industry. Intermodality appears to be the best alternative possible.

References

1. Giffinger, R.: Smart cities ranking of European medium-sized cities. Centre of Regional Science, Vienna University of Technology, October 2007
2. Schwab, K.: The Fourth Industrial Revolution. World Economic Forum, New York (2016)
3. Ryan, L., Turton, H.: Sustainable Automobile Transport: Shaping Climate Change Policy. Economic and Social Research Institute, Chentelham (2007)
4. Camarero, L.A., Oliva, J.: Exploring the social face of urban mobility: daily mobility as part of the social structure in Spain. Int. J. Urban Reg. Res. **32**(2), 344–362 (2008)
5. Urry, J.: Mobilities. Polity Press, Cambridge (2007)
6. Goods, C.: Greening Auto Jobs: A Critical Analysis of the Green Job Solution. Lexington Books, London (2014)
7. Lucas, K.: Transport and social exclusion. Where are we now? Transp. Policy **20**, 105–113 (2012)

Simulation Model of Traffic in Smart Cities for Decision-Making Support: Case Study in Tudela (Navarre, Spain)

Juan-Ignacio Latorre-Biel[1](✉), Javier Faulin[2], Emilio Jiménez[3], and Angel A. Juan[4]

[1] Department of Mechanical, Energy, and Materials Engineering, Public University of Navarre, Campus of Tudela, Tudela, Spain
juanignacio.latorre@unavarra.es
[2] Department of Statistics and OR, Institute of Smart Cities, Public University of Navarre, Pamplona, Spain
javier.faulin@unavarra.es
[3] Department of Electrical Engineering, University of La Rioja, Logroño, Spain
emilio.jimenez@unirioja.es
[4] Department of Computer Science, Open University of Catalonia, Castelldefels, Spain
ajuanp@uoc.es

Abstract. Traffic constitutes a key factor in a city. Thus, city layout, quality of services, pollution, products delivery, people transportation, and many other activities depend closely on traffic. The decision makers of a smart city should conceive ways for limiting pollution, fuel consumption, and transportation times, as well as accidents and disturbances to dwellers, to give a few examples. In order to achieve these objectives, decisions should be made on the appropriate configuration of the reachable degrees of freedom of the traffic system. However, the complexity of traffic systems, and the conflicting goals of the decision makers in a smart city, makes decision support systems a tool to be considered. In this paper one of such systems is described. It is based on the use of a simulation model for supporting the decision making by what-if experiments or by optimization. This model is developed using the paradigm of the Petri nets and is applicable for simulation and for structural analysis. The model is simple and can be easily adapted to different cities or road networks by adding to the model the layout of the city streets and roads, as well as some additional information such as traffic lights or number and type of vehicles.

Keywords: Smart city · Petri nets · Mesoscopic traffic simulation · Decision making support · What-if analysis · Optimization · Smart city management

1 Introduction

Traffic and transportation are concepts with substantial financial, environmental, and social implications on the daily activities of a city. These implications are even more significant for smart cities. Not for nothing, smart cities promote efficiency and sustainability as a way to reduce costs, processing timespan of urban activities, resources, wastes, and emissions, as well as to improve the quality of life of the city dwellers.

© Springer International Publishing AG 2017
E. Alba et al. (Eds.): Smart-CT 2017, LNCS 10268, pp. 144–153, 2017.
DOI: 10.1007/978-3-319-59513-9_15

Decision makers should find ways to achieve these objectives in their sphere of influence. Some of these actors are members of municipal or regional governments with responsibilities on public transportation, emergency systems, or traffic control and regulations, managers of companies with logistic systems, as well as private owners of transportation means. Different technologies provide some help to achieve these goals of efficiency and sustainability, such as sensors, cameras with pattern recognition software, vehicle-to-vehicle (V2V) communications, GPS receivers and navigators, etc.

The confluence of many different factors, such as the cutting-edge devices, large number of actors and transportation means in a city, the strong commercial competition between logistic companies, the increasing expectations of consumers in the quality of products and delivery services, and, in addition, the large number of degrees of freedom, leads to a complex environment, where making efficient decisions is a must but it is not an easy task.

These decisions, when developed on the fly on the real system, can lead to costly financial, environmental, and social implications. Moreover, relying on experienced decision makers may be costly and not always available or even adaptable to a changing environment. In this context, simulation arises as a promising tool for testing decisions before putting them into practice, saving time, financial resources, dwellers' dissatisfaction, or even social unrest [1–3].

Significant effort has been devoted to develop traffic simulators [4], which can be classified according to different criteria, such as:

(a) Open-source or proprietary simulators. The first category allows more freedom to configure simulation parameters and experiment design.
(b) Continuous-time, discrete-time, or discrete-event simulators. The last category may provide with realistic simulations to a certain extent, while consuming limited computer resources. This approach has been followed to develop the simulator described in this document.
(c) Microscopic, mesoscopic, or macroscopic simulators. These categories are associated to a decreasing level of detail in the description of individual vehicles and their behavior. Additionally, less detail, in this case, implies less complexity in the simulator and likely, less computer resources to develop simulations and experiments. The simulator described in this document can be classified as mesoscopic, since it described the vehicles as individual entities but some processes, such as lane changes, are not considered in detail. A mesoscopic simulation is believed by the authors to provide with enough accuracy for the decision problems aimed at, while constraining the computational cost.

A survey on different tools available for traffic simulation is given in [4]. One of the most successful formalisms to represent a discrete event system is Petri nets. A Petri net model presents many useful features, such as a double graphical and matrix-based representation, which allow to represent explicitly the elements and potential evolutions of the system, as well as constitutes a successful way to carry out structural analysis and performance evaluation based on simulation. Petri nets have been applied to the modelling of systems with complex behavior and the development of decision support tools [5–7].

Decision making under uncertainty has been addressed successfully by simheuristics, based on Monte Carlo simulation and metaheuristics [8]. It is possible mixing simheuristics, simulation, and Petri net modeling to deal with decision making under uncertainty applied to logistic systems [9].

As a consequence of all these advantages, Petri nets has been selected to develop the simulation model of a traffic system presented in this document. Most of the previous work in Petri net modelling of traffic is devoted to a very active area of research related to the control of traffic lights in intersections using field data provided by sensors in order to maximize the flow, considering unexpected events, such as accidents [10–14]. They focus on a microscopic modeling, very accurate for small models but requiring large computational resources when dealing with large road or street maps.

Moreover, [15] describe a hybrid Petri net traffic model combining macroscopic modeling in roads and microscopic modeling in intersections. The proposed model presents some common features with the model described in the present document. However, high level Petri nets, such as colored Petri nets (CPN), have not considered, leading to a quite large model for representing a whole community's street map or complex road map.

This document describes a Petri net model of a traffic system for mesoscopic simulation and decision making support, which is in process of being implemented. The main objective of this research is to test the performance of the Petri nets in a decision support system for traffic networks, based on the results, tools, and characteristics of this paradigm, as well as the successful applications of Petri net modelling developed in other sectors. Among the main contributions of the presented model, the Petri net graph and incidence matrix size is not increased as the modelled network grows, since information of the network is read as the model requires it for updating the state of the net in a simulation. Another contribution is the increase in the size of the modelled traffic system, since in most of the references addressing Petri net models of such systems, a microscopic model of a reduced set of crossroads is considered. This use of Petri nets for modelling large traffic networks is also a contribution to general traffic simulators, since many of them are based on programming languages and not in the development of a mathematical model with all the possibilities of this paradigm: structural analysis as well as performance evaluation, integration with tools and models already developed for Petri nets, broad body of knowledge related to Petri nets, possibility of refining the model by top-down modeling, integrating with simheuristics and other approaches for decision making, good insight into the structure of the model by an intuitive graphical representation, open source approach, etc.

The rest of the paper is organized follows: Sect. 2 describes a Petri net model of a traffic network, while Sect. 3 deals with the application of the model for decision making support. Section 4 addresses the conclusions.

2 Petri Net Model for Traffic Simulation in a Smart City. Application Case to a Neighborhood of Tudela (Spain)

CPNs [16] is a formalism that can lead to very compact models in systems such as a traffic system, where many elements differ in features that can be aggregated to the model as attributes of one of its elements: the tokens. CPNs have been applied for performance evaluation and decision making in logistic systems, for example in manufacturing facilities [17]. The Petri net graph of the proposed model of the traffic in a smart city is represented in Fig. 1.

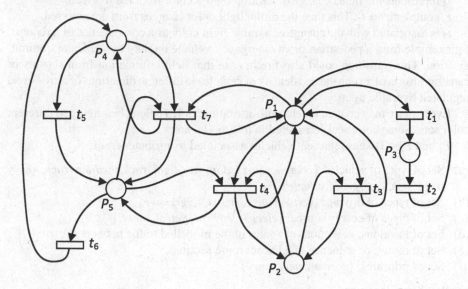

Fig. 1. Petri net model of a traffic system.

Some of the constitutive elements of the model are detailed in the next paragraphs. Firstly, the five places of the net can be described as follows:

P_1 is a place representing a route section, such as a street, street lane, road, or roundabout. A route section is defined as a part of a street, road, or roundabout without intermediate diversions or side-streets. This place can hold tokens representing vehicles of different types.

P_2 describes the capacity of a given route section. A token represents a measure unit of the place, since the place required by a vehicle depends on its type

P_3 contains tokens, representing stopped vehicles. They include parked vehicles as well as damaged or wrecked ones. Notice that a street has also a limited but variable amount of parking places, when considering double parking.

P_4 and P_5 correspond to a red and green traffic light respectively. A token contains information on the location of the traffic light.

The transitions of the Petri net, the second type of nodes, are the following:

t_1 represents the stop of a vehicle in a given route section.

t_2 complements t_1, since it corresponds to the start of a stopped car.

t_3 describes the motion of a vehicle from a route section to an adjacent one. In this case, the connection between both sections is not regulated by a traffic light. A token from the original route section, representing a given vehicle, arrives at the end of its route section, if there is free place in the next section of its route, transition t_3 fires, updating the attribute of the token representing the route section, where the vehicle is moving. The token remains in place P_1.

t_4 is similar to t_3; however, in this case there is a traffic light constraining the traffic from the initial route section to the final one.

t_5 represents the timed change of a traffic light's color from red to green.

t_6 complements t_5. This time the traffic light color changes from green to red.

t_7 is associated with the change of a traffic light color as a consequence of a request, for example from a pedestrian or an emergency vehicle on duty by vehicular communication. This transition could also fire in case that field information from sensors or cameras inform of an unbalanced density of vehicles in different directions of a crossroad regulated by traffic lights.

Every place may contain tokens with attributes, called colors, belonging to different color sets. Some color sets considered in this model are:

P_1 and P_3 – Tokens represent vehicles associated to attributes from:

(a)   Set of types of vehicles: {*compact car, sedan, van, light truck, medium truck, heavy truck, bus, emergency vehicle*}.
(b)   Set of types of driving: {*conservative, smart, aggressive*}.
(c)   Set of types of energy source: {*electric, hybrid, petrol, diesel*}.
(d)   Set of locations, or sections of routes in the modelled traffic network.
(e)   Set of routes, or sequence of adjacent route sections.
(f)   Set of priorities: {*normal, emergency*}.

P_2 – Set of locations.

P_4 and P_5 –Tokens, representing traffic light status, have these color sets:

(a)   Set of identifiers (ID) of traffic lights.
(b)   Sets of time delays for activation of red and green lights.

Some information on the street network should be added to the model. The network of a neighborhood of the town of Tudela, Spain, is shown in Figs. 2 and 4(a), where nodes and route sections are superimposed to the street-map. The model of the street-map comprises 56 nodes (circles), 92 route sections (directed arcs), and 4 traffic lights (triangles).

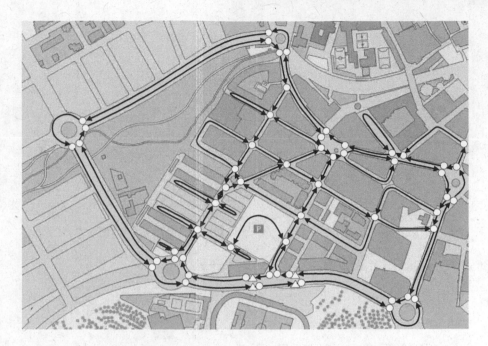

Fig. 2. Street-map of a neighborhood of Tudela with nodes and route sections. (Color figure online)

Figures 3 and 4(b) show elements with their IDs. This area includes roundabouts, one-way and two-way streets, four-lane streets, blind alleys, and parkings.

This information, complementing the Petri net model, is stored in a data structure that is independent to the Petri net model. In fact, the model reads the information it needs to evaluate when transitions t_3 and t_4 are enabled and how their firing modify the attributes of the tokens that change of place.

3   Decision Making Support in Smart Cities with Petri Nets

The previous Petri net model can be integrated in a simulation-based decision-making support tool. A decision can be made by assigning values to the decision variables, parameters of the model, then the evolution of the system can be studied by the simulation of its Petri net model. Due to the fact that the decision variables can be chosen among all the parameters of the Petri net model, the associated decision support system is very flexible, allowing to describe decisions associated to different contexts and decision makers.

The simulation is performed by the so called "token game", describing the dynamic rules of a Petri net evolution. During this simulation one or several performance parameters can be calculated. They can be associated to the decision made as quality parameters. After checking a set of different feasible decisions, their quality parameters can be compared and chosen the best one for the stated problem. This tool can be applied

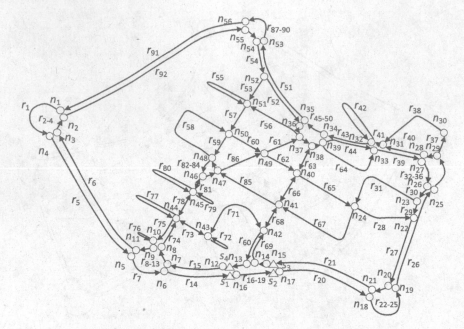

Fig. 3. Traffic network with IDs of nodes, route sections and traffic lights.

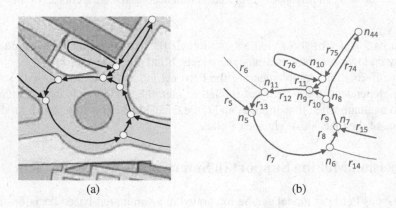

(a)   (b)

Fig. 4. Roundabout network: (a) street map, (b) IDs of nodes and route sections.

for performing experiments under a "what-if" approach or an optimization process, meaning that some modifications are implemented on the system and the simulation's outcome allow to foresee the main consequences of such decision.

The experiments to be performed can be adjusted depending on the scope of the decisions to be made. In other words, the responsibilities of the decision-maker would determine the degrees of freedom of the system. Other parameters of the traffic system would be modeled by means of deterministic or stochastic parameters, meaning that they are uncontrollable or out of the reach of the decision maker. Other experiments can change the degrees of freedom, representing realistically the decision making process.

Examples of decision makers are municipal and traffic authorities, managers of companies with logistic services for delivering products, managers of emergency services or public transportation, as well as private car owners. Some examples of decision makers and several related degrees of freedom of a traffic system are given in the following. All these degrees of freedom in the traffic system are modelled by parameters of the Petri net (time delays associated to transitions, initial markings, priorities in actual conflicts) or in the data structure containing information on the route network:

(a) For traffic authorities it can be considered the activation/deactivation and timing of traffic lights, installing sensors and cameras to adapt the behavior of traffic lights to traffic conditions, closing streets for demonstrations, parades, or works, changing driving senses, building up new streets, roads or roundabouts, pedestrianization of streets, etc.

(b) For logistic planners, routes of every vehicle including the departure time, the amount and types of transported products, number of available vehicles, number and sizes of depots, etc.

In addition to the decision variables it is important to determine the objectives aimed with the decision making process. Some examples are:

(a) For traffic authorities, it can be considered the average driving time of emergency or private vehicles for a certain route, pollution and noise levels, traffic density and average speed at certain periods of time, detection of bottlenecks and deadlocks, etc.

(b) For logistic planners, it can be taken into account the need of meeting the schedules, as well as minimizing the number of vehicles, delivery time, amount of transported products returned to the depot, number of returns to the depot to reload products, petrol consumption, etc.

The degree of achievement of a certain objective, after a decision, can be evaluated by means of performance parameters, which are calculated while simulating the behavior of the Petri net model. For instance, travel times can be calculated by adding the time a given token spend in each route section, modelled by places P_1 and P_3. Congestion and actual capacities of streets can be obtained by adding the marking of P_1 and P_3 (#vehicles) or P_2 (vacancies).

4 Conclusions

The concept of smart city constitute a reference of many communities, where the quality of life of dwellers is a priority. The efficient use of resources and the reduction of wastes and emissions, in short sustainability, are key goals in the management of a smart city and the companies that operate in it. In this context, traffic and transportation activities produce a significant impact in the use of resources, and production of emissions, noise, and wastes.

Decision making should aim at improving the achievement of these objectives. However, the behavior complexity of the system under study and the potential impact of wrong decisions in it, suggest the use of simulation tools to test the feasible decisions in appropriately designed experiments.

The Petri net model for mesoscopic simulation described in this document provides with a flexible tool for decision making support. CPNs lead to a very compact model with only 5 places and 7 transitions. A large variety of individual vehicles, traffic lights, as well as elements of a road and street network can be easily added to the model as attributes of the tokens. This feature provides the model with enough generality to be adapted to any city, town or road network, modeling the behavior of traffic with the level of detail of a mesoscopic simulation. In other words, individual vehicles of interest can be considered, but also deterministic or stochastic macroscopic parameters can be added easily, such as traffic density in streets. An example of application is provided with a neighborhood of Tudela (Spain).

In brief, the proposed Petri net model, presents a compromise between simplicity of the model and level of detail in simulation. It presents potential for the development of flexible decision support tools for many different decision makers with impact in the traffic of the modeled road/street map.

Acknowledgements. This work has been partially supported by the Spanish Ministry of Economy and Competitiveness (TRA2013-48180-C3-P and TRA2015-71883-REDT), FEDER, and the Ibero-American Program for Science and Technology for Development (CYTED2014-515RT0489).

References

1. Latorre-Biel, J.I., Jimenez-Macias, E.: Simulation-based optimization of discrete event systems with alternative structural configurations using distributed computation and the Petri net paradigm. Simul. Trans. Soci. Model. Simul. Int. **89**(11), 1310–1334 (2013). doi:10.1177/0037549713505761
2. Latorre-Biel, J.I., Jiménez-Macías, E., Pérez-de-la-Parte, M.: Simulation-based optimization for the design of discrete event systems modeled by parametric Petri nets. In: Proceedings of the 2011 Fifth UKSim European Symposium on Computer Modeling and Simulation (EMS), pp. 150–155 (2011). doi:10.1109/EMS.2011.63
3. Juan, A.A., Faulin, J., Pérez-Bernabeu, E., Domínguez, O.: Simulation-optimization methods in vehicle routing problems: a literature review and an example. In: Fernández-Izquierdo, M.Á., Muñoz-Torres, M.J., León, R. (eds.) MS 2013. LNBIP, vol. 145, pp. 115–124. Springer, Heidelberg (2013). doi:10.1007/978-3-642-38279-6_13
4. Barceló, J. (ed.): Fundamentals of Traffic Simulation. International Series in Operations Research & Management Science, vol. 145. Springer, Heidelberg (2010). doi: 10.1007/978-1-4419-6142-6
5. Latorre-Biel, J.I., Jiménez-Macías, E., Pérez, M.: Sequence of decisions on discrete event systems modeled by Petri nets with structural alternative configurations. J. Comput. Sci. **5**(3), 387–394 (2013). doi:10.1016/j.jocs.2013.09.001
6. Latorre-Biel, J.I., Jiménez-Macías, E., Pérez-de-la-Parte, M., Sáenz-Díez, J.C., Martínez-Cámara, E., Blanco-Fernández, J.: Compound Petri nets and alternatives aggregation Petri nets: two formalisms for decision-making support. Adv. Mech. Eng. **8**(11) (2016). doi:10.1177/1687814016680516

7. Latorre-Biel, J.I., Jiménez-Macías, E., Blanco-Fernández, J., Martínez-Cámara, E., Sáenz-Díez, J.C., Pérez-Parte, M.: Decision support system, based on the paradigm of the petri nets, for the design and operation of a dairy plant. Int. J. Food Eng. **11**(6), 767–776 (2015). doi:10.1515/ijfe-2015-0063

8. Juan, A., Faulin, J., Grasman, S., Rabe, M., Figueira, G.: A review of simheuristics: extending metaheuristics to deal with stochastic optimization problems. Oper. Res. Perspect. **2**, 62–72 (2015). doi:10.1016/j.orp.2015.03.001

9. Latorre, I., Faulin, J., Juan, A.: Enriching simheuristics with Petri net models: potential applications to logistics and supply chain management. In: Proceedings of the 2016 Winter Simulation Conference, pp. 2475–2486, Washington D.C., USA, 11-14 December (2015). doi:10.1109/WSC.2016.7822286

10. Vazquez, C.R., Sutarto, H.Y., Boel, R., Silva, M.: Hybrid Petri net model of a traffic intersection in an urban network. In: Proceedings of the IEEE International Conference on Control Applications (CCA 2010), pp. 658–664, Yokohama, Japan, September 2010. doi:10.1109/CCA.2010.5611322

11. Soares, M.S., Vrancken, J.: Responsive traffic signals designed with petri nets. In: Proceedings of the IEEE International Conference on Systems, Man and Cybernetics (SMC 2008) (2008). doi:10.1109/ICSMC.2008.4811574

12. Fanti, M.P., Iacobellis, G., Mangini, A.M., Ukovich, W.: Freeway traffic modeling and control in a first-order hybrid petri net framework. IEEE Trans. Autom. Sci. Eng. **11**(1) (2014). doi:10.1109/TASE.2013.2253606

13. Yaqub, O., Li, L.: Modeling and analysis of connected traffic intersections based on modified binary petri nets. Int. J. Veh. Technol. (2013). doi:10.1155/2013/192516

14. Qi, L., Zhou, M.C., Luan, W.: A two-level traffic light control strategy for preventing incident-based urban traffic congestion. IEEE Trans. Intell. Transp. Syst. (2016). doi:10.1109/TITS.2016.2625324

15. Tolba, C., Lefebvre, D., Thomas, P., El Moudni, A.: Continuous and timed Petri nets for the macroscopic and microscopic traffic flow modelling. Simul. Model. Pract. Theory **13**(5), 407–436 (2005). doi:10.1016/j.simpat.2005.01.001

16. Jensen, K., Kristensen, L.M.: Coloured Petri Nets: Modelling and Validation of Concurrent Systems. Springer, Heidelberg (2009). doi:10.1007/b95112

17. Latorre-Biel, J.C., Jiménez-Macías, E., Pérez de la Parte, M., Blanco-Fernández, J., Martínez-Cámara, E.: Control of discrete event systems by means of discrete optimization and disjunctive colored PNS: application to manufacturing facilities. Abstr. Appl. Anal. **2014**(3), 1–16 (2014). doi:10.1155/2014/821707

Virtual Development of a Presence Sensor Network Using 3D Simulations

Rafael Pax[✉], Marlon Cárdenas Bonett[✉], Jorge J. Gómez-Sanz[✉],
and Juan Pavón[✉]

Universidad Complutense de Madrid, Madrid, Spain
{rpax,marlonca,jjgomez,jpavon}@ucm.es
http://grasia.fdi.ucm.es

Abstract. Testing the control and deployment of large networks of sensors and actuators is a complex and expensive task. This paper presents a 3D simulation tool that facilitates testing and measuring this kind of systems in a virtual environment, which alleviates the costs of doing these tasks in a physical setting. This is illustrated with an example of how a presence detection system can be designed to monitor the behavior of a crowd under different stimulus. The simulation does not only involve the devices, but provide input data so that they can be decoupled and analyzed separately. This decoupling allows to experiment with different deployments of sensors while the simulation is still working and evaluate their performance in real time. Such feature can be of assistance for decision making when designing a large installation or improving one. The paper contributes with a proof of concept where a large installation, together with its inhabitants, is simulated. The simulation is used then to create different simulations of photoelectric devices that register the proximity of an individual. The results permit to evaluate networks of such devices and think of different configurations without the limitations of the physical environment and the privacy and integrity concerns of individuals.

Keywords: Smart cities · Simulation · Sensors · Actuators

1   Introduction

There is little support for smart environments design. A relevant issue is the testing of the control and deployment of large networks of sensors and actuators. This is a complex and expensive task when done directly with physical devices and involving people (e.g., actors). There are even some cases where the experimentation is risky (e.g., fire, falls, etc.) This paper presents a 3D simulation tool that facilitates testing and measuring this kind of systems in a virtual environment, which alleviates the costs of doing these tasks in a physical setting.

This paper introduces an architecture for the design of large installations that can be the first step to a large scale simulation of smart cities. The architecture allows to run multiple simulations and evaluate the performance of smart devices

© Springer International Publishing AG 2017
E. Alba et al. (Eds.): Smart-CT 2017, LNCS 10268, pp. 154–163, 2017.
DOI: 10.1007/978-3-319-59513-9_16

connected to it. It is assumed that simulated and real devices offer an API towards other components. In the simulation, simulated devices implement an API in different ways so as to capture what makes each device unique. As a proof of concept, the paper focuses on the creation of a network of presence sensors that satisfies the demands of a client that wants to exhaustively measure the presence in an area.

The case study introduces a simulation that corresponds to different occupations of an area. The simulation has a population of individuals that move through the installation following their daily activities. The client wants to evaluate whether a cheap network of sensors can achieve a quality presence tracking in the installation. The sample sensor network is evaluated and its precision measured.

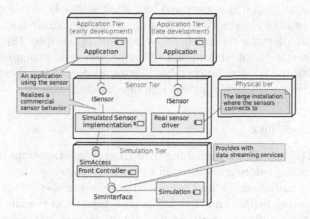

Fig. 1. The different tiers involved in the large installation sensor network development

The organization of the rest of the paper is as follows. Section 2 introduces the architectural solution that allows to run multiple devices. Section 3 is a proof of concept where different simulations of the equivalent of an optical pulse sensor are designed. The related work Sect. 4 shows the connection between this approach and other works. The paper finishes with a conclusions in Sect. 5.

2 Simulation Architecture

The simulation consists of different types of nodes, which are organized in four *tiers*, as shown in Fig. 1: application, sensor, simulation, and physical. All tiers can be executed on different computation nodes, which facilitates the scalability of the model. This can be easily extended as new application and sensor components can be deployed in separate nodes.

This architecture induces a two phases development process, early development and late development. During early development, the work is done against a simulation of the large facility. The simulation provides the input for simulated sensors. These simulated sensors are expected to capture the behavior of real (commercial) ones, faulty behaviors included if needed. Late development assumes the application logic is already or almost built. Then, it can be connected to real sensors and proceed to final adjustments that take into account particularities of the sensors.

For this development cycle to work, the fidelity of the simulated sensor components has to be close enough to real ones, though maybe not exactly the same. Hence, simulated sensors provide the same, or similar, functionality, while real sensors connect to the real world.

Under this assumption, applications can work with both real or simulated sensors. It makes the development easier (and with lower costs) to work initially with simulated sensors and later on evolve to work with real sensors. Once the mission of the application is understood and what sensor layout is required, real deployments can take place with final tuning of the solution to specific contexts.

The tiers architecture reflect this development philosophy through the distinction of three levels of abstraction. This section will explain in detail the sensor, application, and simulation tiers. The physical tier is expected to be the infrastructure that is available for the late development stages.

2.1 Simulation Tier

Previous work [3] has shown that using data from simulations can be very useful for testing purposes before a deployment of devices. However, the computational cost of crowd simulations is highly related to the number of agents (virtual avatars representing people and their behaviors) present in the simulation. Also, the realism of the simulation itself, such as 3D processing, physics or animations increases this cost.

Using this approach might not be affordable for everyone. A good approach for solving this issue is to execute the simulations in a central server, and allowing remote access to the simulation data from lightweight clients, via well-known data-transmission protocols and standards.

The simulation management is done through a front controller that mediates all accesses to the different running simulations. Figure 2 (left) depicts a simulation as an example. The goal is to run in the server as many simulations as required, each one representing particular configurations of the facility. Current version runs the simulations in the same server, but it does not have to be that way. Simulations can be distributed across several computation nodes, which facilitates increasing the number of supported simulations.

A configuration of the simulation includes a 3D description of the physical environment, a population of simulated individuals, and behaviors associated to those individuals. These simulations run in headless mode (i.e. without a GUI) to make a better use of the machine resources. Removing rendering greatly reduces the workload and does not affect the results, since collision detection is still in place despite the absence of rendering.

External clients, in this case, will be the simulated sensors. They will obtain simulation data through a publisher-subscriber way. The client specifies the type of events that need to be sent over the network and then it will just listen to the data stream. For instance, if the sensor being simulated is a people counting sensor placed inside a carpet, 2D data is sufficient. But if the sensor being simulated is a more advanced sensor, such as a motion sensor, the information gathered from the simulation should be more accurate, and 3D information should be provided.

The information supplied can include the physical elements location in the simulation, as well as mobile elements. The information that is currently available from the simulator interface is the position of an element, its orientation, its velocity, mass, current animation (in the case of a person), and the type of element that it is (wall, door, person). The floor map information also provides information about building levels, if any. This selective accuracy and variety of information is relevant to support a higher number of sensor devices.

All information originated in the simulation has a time-stamp. Each event comes with a time-stamp that permits to measure the asynchrony between the data received by the different sensors. Required data bandwidth depends upon the number of elements as well as the required detail level. For the case study in Sect. 3 the required bandwidth per sensor ranges from $3.4\,kB/s$ to $3.7\,kB/s$. The simulation time can be shared across sensors so that they provide a consistent real-time-like vision to the application layer. However, there maybe still a gap between the time as perceived by the application and the time as represented in the simulation. This is something to be improved in new versions of the architecture.

To save bandwidth, a connection made from the client starts with an initial description of the environment. Afterwards, only changes to the initial state are provided. Changes are transmitted over the network labeled as events. Again, the reason is to reduce information overload.

2.2 Sensor Tier

The sensor tier provides real and simulated sensors. Both offer the same interface towards the application, but feed on different data sources. On the one hand, software components representing real sensors are expected to use specific drivers that connect to the physical tier. On the other hand, simulated sensors capture essential features of the real ones and feed on the simulation data output to produce the expected outcome.

The particular implementation of the simulated sensor depends on the needs of the development. It has access to all the simulation events, and it has to process it in a way that the output matches the one expected from a real sensor. Some simulated sensors can include faulty behaviors or use different approaches to reproduce the real sensor behavior.

This decoupling from the simulation is important in order to separate responsibilities in the development. Simulations are already too complex and the inclusion of sensor devices within makes the problem even harder. Also, this allows to test at the same time different configurations or deployments of the same sensor type. Each configuration could be run at the same time and feed on the same data as the rest.

2.3 Application Tier

The real goal of the development is to create the application that uses the sensor data. The development challenge consists in determining how many sensors, and of what kind, are needed to provide the service. To determine this, a fast method is needed that allows to quickly deploy and test sensor networks and evaluate if the provided information is enough to implement the target service. The cost of the sensor network is an important factor as well. There are sensors that provide the exact output that is required, but are too expensive to deploy a network of them. Finding the right combination requires time and experimentation.

The two phases development as proposed in this paper requires the application to assume that there are common interfaces that connect the application to whatever sensor, be it a simulated or a real one. In this way, the application will ignore whether it is connected to the real world or the simulated one.

The performance of the application depends also on the considered scenarios. The decoupling of simulations permits to have different simulations running at the same time, each one capturing crowd simulation scenarios of interest for the development within the simulated facility. Since simulated sensors are connected opportunistically to selected simulations, this allows the application to selectively connect to running simulations and evaluate its performance directly. Besides, this operation would be made in similar conditions as it would be in the real world, since the application does not know if it is connected to real world sensors or not.

Finally, the different tier decoupling allows to run different instances of the same application. This contributes to accelerate the development since all work is not centered in a single computer.

3   Case Study

This practical case deals with the construction of a network of sensors to detects the presence a set of individuals in a physical space, which is simulated by a computer. With this purpose, different presence sensors and configurations are studied.

The sensors can use ultrasonic, optical, inductive, magnetic, and capacitive technology. Those with the longest range are the optical and ultrasonic. Ideally, we want to consider sensors that are allocated only into a single location and do not require a reflector.

Ultrasonic are expensive, but have the longest range (up to 8 m). Price of ultrasonic devices is high (starting from 300 euros a piece). They are not affected by ambient noise or light. Some optical sensors do not require either a reflector and work with an emitter and a receiver that captures reflected signals. In these optical sensors, a pulse is emitted and, if received, the presence sensor is activated. There is some tolerance to ambient light noise and flickers from fluorescents. The drawback is their short range, which is a maximum of 1 m.

The pulse sensor can be modeled as an element that detects objects within a semi-sphere whose center is the sensor and the radius is one meter. The model can be improved, but serves enough for this paper and to evaluate the expected performance of a presence network sensor.

The simulation will start a random stream of people that will enter the facility, who will move through the rooms and corridors randomly. In this case the facility is one of the floors of the building of the Faculty of Computer Science where we work, which is shown in Fig. 2 (left). This figure shows the 3D rendering of the simulation for the sake of clarity only, since, in the simulation architecture as shown in Fig. 1, these simulations usually run in headless mode.

Within the facility, six sensors are located in one of the more traveled corridors, which contains alternate paths or detours to classrooms that will change the movement of individuals.

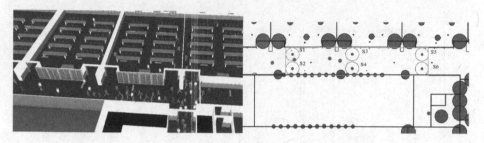

Fig. 2. (Left) 3D rendering of the simulation case study. (Right) Location of sensors in the case study (Color figure online)

The simulated sensors in the sensor tier within this development are provided with a GUI for drawing what information each simulated sensor is accessing. Figure 2 (right) shows this 2D map, which is produced from the 3D simulation run as in Fig. 2 (left). The 2D map shows the x and z axis coordinates of the location pointed out by the mouse. The upper right corner shows the current simulation time. Green circles represent hollows in the walls (windows or doors). Blue dots represent individuals traversing the installation. The red filled circles represent physical objects in the area. Sensors S_1 to S_6 are labeled in the figure at the center of the corridor. The location of the sensor is depicted as a red dot and the range of the sensor, one meter, is drawn as a concentric circle. The controlled area is drawn as an orange rectangle wrapping all the sensors.

The six sensors will be deployed across this orange rectangular region (see Fig. 2 (right)). Each sensor is allocated at the ceiling of the facility and has a range of one meter. The sensor is activated if an individual crosses the hemisphere whose center is the sensor location. The hemisphere that models the range of detection of these sensors will detect those pedestrians who are high enough to be within one meter of the ceiling, when walking underneath. To measure the simulated sensor measurement efficacy, a rectangle covering the corridor is drawn and the people walking across it measured.

The objective is to find the best configuration and distribution of the sensors to build a mesh with a specific number of devices that will allow to detect as accurately as possible the individuals that transit within a certain space. Comparing the sensor detection against the orange rectangle based approach permits to evaluate the precision in the detection.

If the six sensors are numbered from S_0 to S_6, we can model the application that combines their output as Eq. 1. Hence, there will be people in the corridor if any sensor detects anyone.

$$app = S_0 \vee S_1 \vee S_2 \vee S_3 \vee S_4 \vee S_5 \vee S_6 \tag{1}$$

If the function that determines if there is really someone at the corridor is named h, the purpose of the experiment is to assess the suitability of a location and a number of sensors so that the equation $app \approx h$ is satisfied.

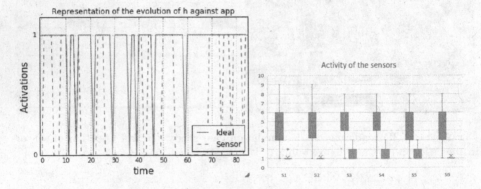

Fig. 3. (Left) Representation of the evolution of h (ideal) against app (sensor). (Right) People present in the proximity of the sensor (blue) vs people present when the simulated sensor triggered (orange). (Color figure online)

The Fig. 3 (right) shows the account of people within the sensor deployment area, the orange rectangle in Fig. 2 (right), but divided into 6 pieces, each one corresponding to the proximity of an individual sensor location (S_1 to S_6). In blue, there is an account of the number of times a person crossed the area assigned to each sensor. In orange, the diagram depicts the number of persons crossing the area when the sensor activated. X axis represent each sensor. Y axis represents the number of persons. It can be seen that the actual number of people crossing the orange rectangle is greater than the number of people triggering the sensors.

There are three hypothesis that explain what is happening. Firstly, the control area is bigger than the range of the sensors. Hence, necessarily, there will be more people crossing the area than people being detected. Secondly, it turns out that some of the individuals did cross the section, but never got close to some sensors. The reason is that these persons crossed doors to enter other rooms and avoiding the sensor location. For instance, a person entering the room close to S_3 may never enter within the S_3 range, but will be accounted by the orange rectangle. Thirdly, characters in the simulation behave following the shortest path. Hence, if characters intend to enter into the room close to S_1, they can enter directly without triggering S_1 and S_2. The sensor S_6 surely is not fired frequently because of the same reason, because the shortest path crosses more times S_5 than S_6.

Despite the low number of people crossing the sensor range area, the total outcome of the presence function is not unsatisfactory. Figure 3 (left) contains the representation of the evolution of *app* and *h* functions. In general, *h* returns more positive activations than *app*. It has already been discussed why this could be, and the reason maybe an unexpected behavior of the pedestrians, that did not traverse the corridor as expected, but got into classrooms and managed to avoid sensors. Nevertheless, the figure does not shows false positives and fails when the pedestrians show this unexpected behavior.

The lesson from this experiment is that the location of the sensors at the ceilings is not a bad idea, but that the behavior of the pedestrians greatly affects the success ration. We think it is possible a better location of the sensors so that better success ratio is obtained.

4   Related Work

The work [2] studies the energy savings in a building based on occupancy sensors. They use acoustic, light, motion, CO_2, temperature and relative humidity sensors. This paper has been proposed with a motion sensor in mind only. It limits then to one kind of sensor and discovering patterns in the data was not the goal. The paper [2] also uses a simulator, energyplus, to test the outcome of the occupancy detection with respect the energy consumption in the building. In this work, the simulation is used to generate data for the sensors, and no actuator is defined.

The paper [6] introduces the Tokyo Virtual Living Lab for conducting controlled driving and travel studies. The goal is to raise awareness in order to create an eco-sustainable and optimized transportation. The participation of users in the simulation allows to foresee the effect of traffic and experiment with different ways of driving styles. This paper does something similar, since it intends to use a 3D simulation as testbed for testing ideas. However, this paper focuses on devices and does not involve human participation.

The work [4] point at 3D as a useful tool to support urban operational decision makers. The work uses 3D models to represent the outcome of some natural, such as floodings, or artificial events, energy outages. The system feeds from real data to provide a mixed reality perspective. Our approach is different in that the simulation is not fed with external data, but it creates the different data streams through the aim of game engines. Also, it facilitates the development of devices. Simulated devices are connected to the simulation and they can be used to service other applications. Then, the applications, if the interface to the simulated sensors is sufficiently accurate, can switch the kind of sensors used and connect to real ones, if available. In this way, the same application can be reused in a different context.

The work [1] uses Ubiksim to simulate a Smart Campus and study how an agent based social simulation model can help and influence decision making processes in the context of the faculties management. The differences with the current proposal is that Ubiksim is based on 2D engines that do not account for three dimensions. Hence, situating a device in the ceiling or at some height is not explicitly accounted. Also, the height of the characters or peculiarities in their body movement are not captured, so this particular problem where sensors detect presence within the detection area of a sensor (represented by a cone) would not have been possible.

[7] introduces a tool chain where the environment is reproduced through a 3D simulation created with SweetHome3D software, as in Ubiksim. Then, they use a middle-ware based on ROS (Robot Operating System) that decouples the application from the simulation and permits to switch between real sensors and simulated sensors. The physical reactions in the simulation are achieved through a plugin called Gazebo. The authors have to determine some consequences due to the use of limited physics. This approach is similar to ours. In fact, early work was made with SweetHome3D [5] and cannot achieve the performance required for thousands of characters. Besides, SweetHome3D does not include physics either, and that forced authors to use Gazebo. We have solved this by using a game engine, JMonkey, and a physics engine, Bullet. These elements may suggest that this approach is more suitable towards the simulation of large spaces.

5   Conclusions

The paper has introduced an architecture that enables the simulation of the equivalent of optical presence sensors in a large installation. The network is used to measure the presence in a large installation through cheap sensors that are opportunistically distributed. This allows to choose which areas and with what precision presence has to be detected beforehand. The paper has shown how such network can be simulated against different crowd simulation configurations. Also, how the performance of the simulated devices and their efficacy can be measured to imply how well the actual deployment works. It has shown as well the importance of including crowd simulation elements that permit to test the idea in working conditions closer to real ones. In particular, it served to identify issues with an intuitive six sensor deployment across a corridor.

The paper has not addressed which sensor behavior is more suited towards this task. It is assumed that the developer will reproduce the behavior of different commercial sensors and translate them to the simulation. Then, a suitable combination of them can be used to solve the problem or, at least, to know more.

The size of the crowd simulation is a modest one of some hundreds of individuals. More experiments have been made with thousands, but it is still pending work to achieve magnitudes of populations like those to be found in cities.

Acknowledgements. We acknowledge support from the project "Collaborative Ambient Assisted Living Design (ColoSAAL)" (TIN2014-57028-R) funded by Spanish Ministry for Economy and Competitiveness; and MOSI-AGIL-CM (S2013/ICE-3019) co-funded by Madrid Government, EU Structural Funds FSE, and FEDER.

References

1. Campuzano, F., Doumanis, I., Smith, S., Botia, J.A.: Intelligent environments simulations, towards a smart campus. In: 2nd International Workshop on Smart University (2014)
2. Dong, B., Andrews, B.: Sensor-based occupancy behavioral pattern recognition for energy and comfort management in intelligent buildings. In: Proceedings of building simulation, pp. 1444–1451 (2009)
3. Gómez-Sanz, J.J., Cárdenas, M., Pax, R., Campillo, P.: Building prototypes through 3D simulations. In: Demazeau, Y., Ito, T., Bajo, J., Escalona, M.J. (eds.) PAAMS 2016. LNCS (LNAI), vol. 9662, pp. 299–301. Springer, Cham (2016)
4. Howell, S., Hippolyte, J.L., Jayan, B., Reynolds, J., Rezgui, Y.: Web-based 3D urban decision support through intelligent and interoperable services. In: 2016 IEEE International Smart Cities Conference (ISC2), pp. 1–4. IEEE (2016)
5. Pax, R., Pavón, J.: Agent-based simulation of crowds in indoor scenarios. In: Novais, P., Camacho, D., Analide, C., El Fallah Seghrouchni, A., Badica, C. (eds.) Intelligent Distributed Computing IX. SCI, vol. 616, pp. 121–130. Springer, Cham (2016). doi:10.1007/978-3-319-25017-5_12
6. Prendinger, H., Gajananan, K., Zaki, A.B., Fares, A., Molenaar, R., Urbano, D., van Lint, H., Gomaa, W.: Tokyo virtual living lab: designing smart cities based on the 3D internet. IEEE Internet Comput. **17**(6), 30–38 (2013)
7. Roalter, L., Moller, A., Diewald, S., Kranz, M.: Developing intelligent environments: a development tool chain for creation, testing and simulation of smart and intelligent environments. In: 2011 7th International Conference on Intelligent Environments (IE), pp. 214–221. IEEE (2011)

Author Index

Addouche, Sid-Ali   86
Alawadi, Sadi   29
Alba, Enrique   51, 107, 128
Alvarez, Pablo   39
Álvarez-García, Juan Antonio   63
Álvarez-Socarrás, Ada M.   11
Arellano-Arriaga, Nancy A.   11

Ballon, Pieter   97
Bastidas, Viviana   20
Bezbradica, Marija   20

Cárdenas Bonett, Marlon   154
Castillo, Pedro A.   75
Chicano, Francisco   128
Cintrano, Christian   128

Dellagi, Sofiene   86
Déniz, Oscar   63

El Mhamedi, Abderrahman   86
Enríquez, Fernando   63

Faulin, Javier   39, 144
Fernández-Ares, Antonio   75
Fernández-Delgado, Manuel   29

Gachet, Diego   1
Garcia-Arenas, Maria   75
Gómez-Sanz, Jorge J.   154

Haasdijk, Evert   118
Helfert, Markus   20

Jiménez, Emilio   144
Juan, Angel A.   144

Latorre-Biel, Juan-Ignacio   144
Lerga, Iosu   39
Lopatnikov, Daniel   137

Martínez-Salazar, Iris A.   11
Mera, David   29
Merelo, Juan J.   75
Morell, J.A.   51

Pavón, Juan   154
Pax, Rafael   154

Sasián, Félix   1
Serrano, Adrian   39
Smit, S.K.   118
Soria, Luis Miguel   63
Stolfi, Daniel H.   107
Suffo, Miguel   1

Taboada, José A.   29
Therón, Ricardo   1
Troudi, Asma   86

Velasco, Francisco   63

Walravens, Nils   97

Yao, Xin   107

Zonta, Alessandro   118

Printed in the United States
By Bookmasters